Introduction to Virtual Reality

Introduction to Virtual Reality

Alvin Albuero De Luna

ARCLER
PRESS

www.arclerpress.com

Introduction to Virtual Reality

Alvin Albuero De Luna

Arcler Press
224 Shoreacres Road
Burlington, ON L7L 2H2
Canada
www.arclerpress.com
Email: orders@arclereducation.com

ISBN: 978-1-77469-141-0 (Hardcover)

ABOUT THE AUTHOR

Alvin Albuero De Luna is an instructor at a Premier University in the Province of Laguna, Philippines - the Laguna State Polytechnic University (LSPU). He finished his Bachelor's degree in Information Technology at STI College and took his Master of Science in Information Technology at LSPU. He isnhandling Programming Languages, Cyber Security, Discrete Mathematics, CAD, and other Computer related courses under the College of Computer Studies

TABLE OF CONTENTS

LIST OF GLOSSARY

A

Amend – make minor changes to (a text, piece of legislation, etc.), in order to make it fairer or more accurate, or to reflect changing circumstances.
Anatomy – the branch of science concerned with the bodily structure of humans, animals, and other living organisms, especially as revealed by dissection and the separation of parts.
Anomaly – something that deviates from what is standard, normal, or expected.
Anxiety – It is a normal and often healthy emotion. However, when a person regularly feels disproportionate levels of anxiety, it might become a medical disorder.
Artificial Intelligence – refers to the simulation of human intelligence in machines that are programed to think like humans and mimic their actions.
Authentic – of undisputed origin and not a copy; genuine.

B

Bandwagon – used in reference to an activity, cause, etc., that is currently fashionable or popular and attracting increasing support.
Baristas – a person who serves in a coffee bar.
Betrayals – the action of betraying one's country, a group, or a person; treachery.
Bout – A short period of intense activity of a specified kind.

C

Chemotherapy – The treatment of disease by the use of chemical substances, especially the treatment of cancer by cytotoxic and other drugs.
Coherence – the quality of being logical and consistent.
Concurrently – at the same time; simultaneously.
Constituent – being a part of a whole.
Contingencies – a future event or circumstance which is possible but cannot be predicted with certainty.
Conventional – based on or in accordance with what is generally done or believed.
Conveying – transport or carry to a place.
Cyberspace – the notional environment in which communication over computer networks occurs.

D

Dampened – To check or diminish the activity or vigor of.

E

Elucidate – make (something) clear; explain.
Embracing – hold (someone) closely in one's arms, especially as a sign of affection.
Eminent – (of a person) famous and respected within a particular sphere.

Ergonomics – the study of people's efficiency in their working environment.
Evolution – the process by which different kinds of living organism are believed to have developed from earlier forms during the history of the earth.

F
Feasible – possible to do easily or conveniently.

G
Gesture – A gesture is a form of non-verbal communication or non-vocal communication in which visible bodily actions communicate particular messages, either in place of or in conjunction with speech. Gestures include movement of the hands, face, or other parts of the body.
Graphics – the use of diagrams in calculation and design.

H
Hallucination – are sensations that appear to be real but are created within the mind.
Haptic – Haptic technology, also known as kinesthetic communication or 3D touch, refers to any technology that can create an experience of touch by applying forces, vibrations, or motions to the user.

I
Immersive – Generating a three-dimensional image that appears to surround the user.
Imperceptible – so slight, gradual, or subtle as not to be perceived.

K
Kinesthetic – relating to a person's awareness of the position and movement of the parts of the body by means of sensory organs (proprioceptors) in the muscles and joints.

L
Latency –from a general point of view is a time delay between the cause and the effect of some physical change in the system being observed, but, known within gaming circles as "lag," latency is a time.
Lucidity – clarity of expression; intelligibility.

M
Monoscopic – that may be viewed using only one eye at a time.
Multifarious – having many varied parts or aspects.

O
Optimistically – in a way that shows hope and confidence about the future.
Orientation – the action of orienting someone or something relative to the points of a compass or other specified position.
Overlain – lie on top of.

P

Perception – It is the organization, identification, and interpretation of sensory information in order to represent and understand the presented information or environment.

Phobias – an extreme or irrational fear of or aversion to something.

Polygon – In geometry, a polygon can be defined as a flat or plane, two-dimensional closed shape with straight sides. It does not have curved sides.

Projection – is the process of converting a 3D object into a 2D object.

R

Resolution – is the number of pixels (individual points of color) contained on a display monitor, expressed in terms of the number of pixels on the horizontal axis and the number on the vertical axis.

Revolutionizing – change (something) radically or fundamentally.

Robotics – is an interdisciplinary field that integrates computer science and engineering. Robotics involves the design, construction, operation, and use of robots.

S

Semantics – the branch of linguistics and logic concerned with meaning. The two main areas are logical semantics, concerned with matters such as sense and reference and presupposition and implication, and lexical semantics, concerned with the analysis of word meanings and relations between them.

Simulation – A simulation is an approximate imitation of the operation of a process or system that represents its operation over time. Simulation is used in many contexts, such as simulation of technology for performance tuning or optimizing, safety engineering, testing, training, education, and video games.

Software – the programs and other operating information used by a computer.

Spatial – relating to or occupying space.

Stereoscopic – Stereoscopy, sometimes called stereoscopic imaging, is a technique used to enable a three-dimensional effect, adding an illusion of depth to a flat image.

Symposium – a conference or meeting to discuss a particular subject.

Synthesis – The combination of components or elements to form a connected whole.

T

Traverse – travel across or through.

U

Universality – the quality of involving or being shared by all people or things in the world or in a particular group.

Utterance – a spoken word, statement, or vocal sound.

LIST OF FIGURES

LIST OF TABLES

LIST OF ABBREVIATIONS

2D	Two-Dimensional
3D	Three-Dimensional
AI	Artificial Intelligence
AR	Augmented Reality
ATM	Asynchronous Transfer Mode
B-ISDN	Broadband Integrated Services Digital Network
BOOM	Binocular Omni-Orientation Monitor
CAD	Computer-Aided Design
CAE	Computer-Aided Engineering
CAID	Computer-Aided Industrial Design
CG	Computer Graphics
CGI	Computer Generated Imagery
CRT	Cathode Ray Tubes
DoF	Degrees of Freedom
EIAS	Electric Image Animation System
EPSRC	Engineering and Physical Sciences Research Council
FOV	Field of View
GPU	Graphic Processing Unit
HCI	Human-Computer Interaction
HKMA	Hong Kong Monetary Authority
HMD	Head-Mounted Displays
HMDs	Head Mounted Display Systems
HRTF	Head-Related Transfer Function
HUD	Heads Up Displays
IO	Input-Output
ILD	Interaural Level Difference
IMU’s	Inertial Measurement Units

ITD	Interaural Time Difference
LCD	Liquid Crystal Display
LCS	Liquid Crystal Shutter
MIS	Minimally Invasive Surgery
MoCap	Motion Capture
MR	Mixed Reality
OHMD's	Optical Head-Mounted Displays
OLED	Organic Light Emitting Diode
PC	Personal Computer
PS4	Play Station 4
Psi	Plausibility
RCSI	Royal College of Surgeons in Ireland
SADIE	Spatial Audio for Domestic Interactive Entertainment Project
SDK	Software Development Kit
SIVR	Semi-Immersive VR
SVG	Scalable Vector Graphics
UI	User Interfaces
UNC	University of North Carolina
UNIs	User Network Interfaces
VCASS	Visually Coupled Airborne Systems Simulator
VCS	Visually Coupled System
VE	Virtual Environment
VIEW	Virtual Environment Workstation Project
VPC	Permanent Virtual Connections
VPN	Virtual Private Network
VR	Virtual Reality
VRML	Virtual Reality Modeling Language
VSync	Vertical Sync
VWG	Virtual World Generator
WIMP	Window, Icon. Menu, Pointer
WoW	Window on World

PREFACE

Virtual reality (VR) is a computer-generated simulation in which people can easily interact within an artificial 3D environment. The development of virtual reality is intended to give people a realistic experience inside a virtual world, by making use of technologies that influence the human brain into believing whatever they are experiencing is real.

Most commonly, virtual reality is used for entertainment purposes such as 3D cinema and video games. With the COVID-19 restrictions in the year 2020, VR is experiencing a huge rise. As per Grand View Research, the global VR market will grow to 62.1 billion dollars in the year 2027.

This book takes the readers through the brief introduction about virtual reality, focuses on history of virtual reality, types of virtual reality and its development, systems of VR, its comparison with augmented reality (AR), human beings in virtual environment (VE), benefits, and applications of VR along with its future.

The first chapter sheds light on the basic overview of the virtual reality so that the readers are clear about the fundamentals and history behind that form the utmost basics in the field. This chapter will also underline the development of virtual reality across the several years, and also explains certain pros and cons related to it.

The second chapter stresses on different systems related to virtual reality. This chapter takes the readers through different types and components of VR systems. It also focuses on comparison between immersive and non-immersive VR systems. Further, the third chapter explains visual, acoustic, and haptic perception in VR along with the key developments in VR haptics.

Then, the fourth chapter compares virtual reality with augmented reality. In the chapter, different components and devices of augmented reality are discussed. Further, certain applications, advantages, and disadvantages are explained towards the end.

The fifth chapter further focuses on how human beings experience in virtual environment. It explains the impact of VR technology on human lives and the efficiency of human performance in virtual world, along with the positives and negatives of escaping to virtual reality.

The sixth chapter explains different technologies in virtual reality, tracking of different gestures. The chapter also discusses binaural audio for virtual and augmented reality and comparing touch-based and head-tracking navigation techniques in a virtual reality. Then, the seventh chapter focuses on several benefits and applications associated with virtual reality in various fields or sectors such as healthcare, talent development, tourism, etc.

In the last chapter, this book emphasizes on the future of virtual reality in different fields. Also, this chapter explains future challenges associated with virtual reality and its development.

This book has been designed to suit the knowledge as well as pursuit of the researcher and scholars and also, to empower them with different aspects of virtual reality, the development of virtual reality and its applications in different sectors, so that they are updated with the information.

CHAPTER 1

Virtual Reality: An Overview

CONTENTS

Virtual reality (VR) is defined as a computer-generated simulation in which people can easily interact within an artificial three-dimensional (3D) environment by using certain electronic devices. The experience can be similar to or completely different from the real world. The chapter initially deals with the basic overview of VR and along with its detailed history. Further, four types of VR are discussed followed by its development and how actually it works. Towards the end of the chapter, certain benefits, limitations, and challenges are discussed to get a clear view of VR.

1.1. INTRODUCTION

Advance fields of engineering, design, entertainment, medicine, education, and training have virtual reality (VR) applied to it in the current times. The real world is mimicked by VR which is a computer interface so that an immersive three dimensional (3D) visual experience is given beyond the flat monitor.

In static, two dimensional (2D) images, the reconstruction of distances and scales between objects is hard. Depth is brought about in objects through three dimensions (Figure 1.1).

Figure 1.1: Virtual reality mimics the real world and gives an immersive 3-D experience.

Source: Image by Springwise.

Artificial intelligence (AI) and computer graphics (CG) involved in Computer technology help create a simulated environment with which humans interact, and this interaction in VR. Literally speaking reality means truth and virtual means imaginary. The animated objects and virtual objects can be actually held by humans in VR.

VR created 3D environment has numerous applications which are quite interesting, e.g., computer games, for driving practice of airplanes and cars along with many other such uses. Augmented reality (AR) is often confused with VR.

In AR, virtual objects are mixed in a real environment, e.g., when Snapchat has stickers added to it. In VR, humans are introduced to a simulated 3D environment so that an illusion is created where they undergo an experience of the 3D scene's sight and sound so that the illusion seems real. For instance, in the movie 'Ra' the protagonist created a game so that the child interacted with R through a VR headset, whereas the virtual environment (VE) was the other gear.

Interesting Facts about VR:

- The first VR headset was patented in the 1960's and was named the 'Telesphere Mask' by inventor Morton Heilig.
- The PlayStation VR headset developed by Sony engineers came about from an open-ended tinkering session without any executive direction.
- Virtual Reality Travel is a thing. In 2015, Marriott boasted about its ability to transport clients from London to Maui in 90 seconds, thanks to Oculus.

The world of CG is now easily accessible to even the average user. The deep interest in VR that commences with computer games becomes a lifelong fascination. The world that has not yet been created becomes available, the access to experiences and surroundings which is not fathomable in real life becomes possible.

There are neither any constraints nor borders to this world of three-dimensional (3D) graphics so that people can use the 4th dimension, namely their imagination to create and manipulate it as per their requirement. However, humans being the way they are, want to go a step further and rather than just watching things on a monitor, want to interact with this world by stepping into it.

VR has become increasingly fashionable and popular in the current decade. Even though the public started noticing VR in the late 1980s and 1990s, an interest was shown in this field during the 1950s. Jaron Lanier, a pioneering computer scientist holds the credit for introducing the term 'VR' to the world in 1987.

Its profile was raised by the movies like The Lawnmower Man even as the 1990s continued to see a lot of research in this field. Most of the times, a computer screen is used to display the VR experiences or special stereoscopic displays are used as these experiences are essentially visual. All the same headphones or speakers may be used for the auditory stimulation that may be involved in VR.

A wired glove, a mouse or a keyboard may be used to interact with the VE. The history of VR is essentially a history of attempts to make the experience as real as possible. Most of the attempts in the history are based on visual experiences are the auditory ones are far lesser.

Hearing is the second sense after visual that imparts the maximum amount of information, and this is the basis for the historical examples in this field. Visual or auditory perception accounts for about 90% of what we feel about the world.

1.1.1. Some Basic Definitions and Terminology

The computer community uses VE and VR interchangeably. There are a number of other terms as well, even though these two are the most used and most popular terms. A few of the important ones are: Virtual Worlds, Artificial Reality, Synthetic Experience and Artificial Worlds. The meaning of all these names is essentially the same.

3D models are used in real-time interactive graphics to which a display technology is combined so that the user gets immersed indirect manipulation and the model world within what is called the VEs.

Virtual has been defined as "being in effect but not in actual fact" where environment stands for "the conditions, circumstances, and influences surrounding and affecting an organism" as per the Ninth Edition of the Merriam-Webster's New Collegiate Dictionary.

Relying on a 3D hand/body tracking, binaural sound and stereoscopic head-tracker displays VR creates an illusion whereby rather than thinking that the environment is being observed externally, the participant feels that he/she is participating in the synthetic environment. VR essentially gives a multi-sensory and immersive experience.

Devices like the DataGlow and 3D graphics used in computer simulations enable interaction of the user with the simulation. The combination of technologies and the computer-generated environments by which an environment is created which is immersive, multisensory, interactive, three dimensional and viewer centered is referred to as VR.

In a way, VR is a clone of the actual physical reality whereby individuals can view and navigate a 3D world in real-time.

A research associate at the University of Washington, Jerry Prothero works in the Human Interface Technology Laboratory and he defines virtual reality by stating that: "*It can be defined in technological terms as a set of input devices which stimulate a high percentage of our sensory input channels, for instance, by providing a wide visual field-of-view and stereo sound. It can be defined in psychological terms a pattern of sensory stimuli which gives one an impression of being in a computer-generated space.*"

All these definitions are the same in essence, even though they may be slightly different from each other. They all imply that through VR in a simulated or autonomous world, an immersive and interactive experience can be had where one actually feels being present in the environment.

Instead of VR, VEs is the term used by a number of people like the researchers mainly due to the unrealistic expectations and hype that are associated with it.

Furthermore, whilst mentioning VR there must be a mention of two other terms which are quite important, namely: Cyberspace and Telepresence. These terms have a context that slightly different from VU even though they are often coupled with it.

- ***Telepresence:*** Marvin Minsky coined the term in 1980 while referring to the teleoperation systems used for controlling physical objects remotely. In this VR of a specific kind, a real environment is simulated remotely in terms of scale or distance.

Precisely speaking telepresence is said to occur when there is enough dexterity with the manipulators at the work site so that at the control station, the operator is allowed to perform normal human functions and a feeling of being actually present as the worksite is generated by providing the operatory with sensory feedback that has sufficient quality and quantity.

- ***Cyberspace:*** William Gibson not only invented the term but also defined it as "a consensual hallucination experienced daily by billions of legitimate operators (…) a graphics representation of data abstracted from the banks of every computer in human system." In the current times, Cyberspace finds association with the World Wide Web (Internet) and entertainment systems.

- ***Telexistence:*** In 1980 and 1981, Susumu Tachi first proposed the concept of telexistence as patents and in 1982, the first report was published in Japanese, followed by the English one in 1984. In this, a human being is enabled from having a sensation that is real-time about being present in a place other than the one where she or he is actually present wherein interaction with the remote environment is possible where the environment may be a combination of the virtual and real or either of them.

An advanced kind of teleoperation system too is referred to by telexistence. Here the operator feels that he/she exists in a surrogate robot which is working in an environment that is remote as he/she is able to perform the remote tasks dexterously by sitting at the control.

- ***Human-Computer Interaction (HCI):*** Humans are able to interact with computers through the process and study of HCI. In advanced HCI, the interactions between the computer and the person can be thought-controlled whereas a mouse and a keyboard can be the interface in the basic HCI.
- ***Haptics:*** When touch is used to sense a synthetic or natural mechanical environment, the capability to do so is referred to as haptics. The ability to perceive one's movement, body position and weight or kinesthesia is also included in haptics.
- ***Haptics Technologies:*** A computer represents certain movements and physical properties of the virtual objects, and a force feedback is provided to the users through haptics technologies. For instance, based on a video game's actions, dynamic resistance is offered by a haptic joystick.

Both the kinesthetic or motion and touch elements are incorporated in haptics. The user can feel the stimulated real physical properties like momentum, texture, weight, resistance, or friction through the interfaces of haptics so that the happenings on the screen are communicated to the user.

1.2. HISTORY-TIMELINE OF EVENTS AND TECH DEVELOPMENT

A science-based fiction story, Pygmalion's Spectacles was released by Stanley in 1935. The main character in the book gets transported to a fictional world through the glasses he wears so that his senses get aptly stimulated and holographic recordings are featured.

In a way, this story predicted what achievements and aims were likely to be entailed with respect to VR in the future so that the story started getting considered as the concept's origin. Nevertheless, it was in the 1830s that the first developments in VR started to become visible.

- **1838:** The Royal Medal of the Royal Society was awarded to Sir Charles Wheatstone in 1840 for describing stereopsis for the first time in 1838 to explain binocular vision whereby owing to this research, he was able to construct the stereoscope.

It was demonstrated in the research that a sense of immersion and depth (3-dimensional) can be given to an image when two photographs (one eye viewing each) taken from different sides of the same object are combined by the brain.

Stereoscope's earliest type could be created by Wheatstone using this technology. A pair of mirrors placed at an angle of 45° to the eyes of the user was used in this where each mirror reflected a picture that was located off to the side.

- **1935:** In the short story Pygmalion's Spectacles a fictional model of VR was presented by Stanley Weinbaum, an American science fiction writer in the year 1935. A pair of goggles was invented by a professor in the story who meets the main protagonist enabling him to see a movie that gave taste, touch, sight, smell, and sound so that whilst in the story "you speak to the shadows (characters) and they reply [...] the story is all about you, and you are in it."
- **1956:** The first VR machine that was patented in 1962, the Sensorama was created by Morton Heilig who was a cinematographer. Up to four people could be accommodated in a large booth which was the Sensorama. All the senses were stimulated by it through a combination of multiple technologies: it had a combined audio, full color 3D video, smell, vibrations, and atmospheric effects like wind.

A scent producer, stereoscopic 3D screen, stereo speakers and vibrating chair were used to give these effects. Wanting the people to fully immerse themselves in films they watched, Sensorama was considered by Heilig as the "cinema of the future." For this, six short films were created.

- **1960:** The first head-mounted display (HMD), the Telesphere Mask was also patented by Heilig. Stereo sound and a wide vision accompanied the stereoscopic 3D images that this mask projected. At this point no motion tracking was present in the headset.

- **1961:** In 1961 two Philco Corporation engineers Bryan and Comeau created Headsight. A tracking system and video screens for each eye were built in this. All the same, this was used only in the military where it allowed them to look at hazardous situations remotely, and it was not available for VR. To enable the user to look around the setting, the head movements were imitated by a remote camera.
- **1965:** A computer scientist, Ivan Sutherland presented his view of the Ultimate Display. Viewed through an HMD, the virtual world was recreated in a manner that it was so real that it would not be possible for the viewer to distinguish it from the actual reality.

Eventually, a computer kept in a room would be able to control the existence of matter whereby the room would have a chair for the seating of the user. In such a room, it would be fatal to display a bullet and confirming if handcuffs were displayed. If programmed well, the display could actually be the Alice's wonderland.

- **1966:** Air Force's first flight simulator was created by a military engineer, Thomas Furness. Thereafter a lot of funding was done by the military so that better flight simulators could be produced, thereby assisting in VR's progression.
- **1968:** The Sword of Damocles, the first VR HMD was created by Sutherland and his student Bob Sproull. This showed only those virtual wire-frame shapes that were simple and instead of connecting to a camera, this head-mount was connected to a computer.

Due to the tracking system every time the users moved their head the perspective was changed by these 3D models. However, the users had to be strapped into this set up as it was suspended from the ceiling, making it quite uncomfortable for them to wear this heavy equipment as a result of which it stayed just a lab project and could not be developed further.

- **1969:** Using video systems and computers, a succession of "artificial reality" was developed by a computer artist-Myron Krueger. Environments that were computer-generated were created by him, and these responded to the people in that environment. VIDEOPLACE technology that finds a mention later was as a result of these projects.

- **1972:** A computerized flight simulator was built by General Electric Corporation wherein with the usage of three screens which surrounded the cockpit, a 180° field of vision was featured.
- **1975:** The Milwaukee Art Centre displayed the first interactive VR platform namely Krueger's VIDEOPLACE. It did not use gloves or goggles, whereas position-sensing technology, video cameras, CG, video displays and projectors were used. In VIDEOPLACE large video screens surrounded the user in VR in dark rooms.

The movements of the users were transferred onto the silhouette after they were recorded so that the users could see their own movements being imitated by the computer-generated silhouettes. Thus, the idea that even when the people were not close physically, communication within a virtual world was possibly got a boost.

- **1977:** MIT created the Aspen Movie Map. Users could virtually wander through Colorado's Aspen city using this program something like with Google Street View. Polygons, winter, and summer were its three modes. A car would drove through the city and captured photographs which were used to create this. It suggested that people could be transported virtually to other places as it entailed first-person interactivity even though no HMDs were involved.
- **1979:** For military purposes, VR was integrated into the VITAL helmet, the HMD by McDonnell-Douglas Corporation. The pilot's eye movements were followed by a head tracker in HMD which matched the images generated by the computer.
- **1980:** Stereo vision glasses were created by Stereographic.
- **1982:** Defanti and Sandin created the Sayre gloves, the first wired gloves. The gloves' fingers had photocells and light emitters which helped monitor the hand movements. The amount of light that hit the photocell varied when the fingers were moved by the user, and these movements of the fingers were then converted into electrical signals. The beginnings of gesture recognition were probably laid here.

A virtual flight simulator's working model called the visually coupled airborne systems simulator (VCASS) was created for the military by Furness.

- **1985:** VPL Research, Inc was founded by Thomas Zimmerman and Jaron Lanier. Amongst all the companies, VR gloves and goggles are known to have been first made by this company. A range of VR equipment was developed by them like the EyePhone HMD, Audio Sphere and DataGLove.
- **1986:** A flight simulator better known as the Super cockpit was developed by Furness between 1986 and 1989. The main features of the training cockpit were: computer-generated #d maps, advanced radar, and infrared imagery whereby the pilot could in real time hear and see.

The aircraft could be controlled by the pilot using gestures, eye movements, and speech through the sensors and tracking system of the helmet (more can be read about Thomas Furness).

- **1987:** Featuring speech recognition, the Virtual Cockpit was developed by the British Aerospace where similar to Furness' Super Cockpit HMD was used. The term "virtual reality" was popularized by Jaron Lanier whilst he was at VPL Research. Sun Microsystems later bought the patents with respect to VR and graphics. A software was built by the company Dimension International whereby 3D worlds could be built in a PC.
- **1989:** NASA gave a contract to Scott Foster to develop a VR training simulator for astronauts-the virtual environment workstation project (VIEW) and he founded the Crystal River Engineering Inc on receiving this contract. Real-time binaural 3D audio processing was developed by this company.
- **1990:** At London's CG 90 exhibition, a VR arcade machine, the Virtuality was exhibited by Jonathan Waldern.
- **1991:** In spite of the signal delays that were likely to take place between the Earth and Mars, A VR system was designed by a NASA scientist – Antonio Medina do that the robot rovers could be driven in real-time to Mars. Computer Simulated Teleoperation was the name given to this system.

Virtuality was launched by the Virtuality Group. A 3D gaming world was available to the players for playing games in the VR arcade machines which were the first VR entertainment system to be mass-produced.

Real-time immersive stereoscopic 3D images and VD headsets were featured in a Virtuality pod. For multi-player games, it was possible to network together some of the machines. In the end, VR versions were there in many popular games like the Pac-Man.

SEGA made an announcement that for the purchase by the general public, they were working on the SEGA VR headset. The Mega Drive console and arcade games would be able to use this headset. Being influenced by movies like RoboCop, this had a visor-like look. For tracking the head movement, the visor had LCD displays placed on it along with sensors and stereo headphones.

Despite the fact that four games were made for it, this was not released. It was considered that the VR effect was quite realistic, and this may cause people to injure themselves, and due to this main concern, it was never released. All the same due to the processing power being limited, this did not really seem to have possible.

- **1994:** A motion simulator arcade machine, the SEGA VR-1 was released by SEGA. CyberMaxx, a VR headset was released by VictorMaxx.
- **1995:** A 3D monochrome video games playing console called the Visual Boy was launched by Nintendo. Thus, 3D graphics were for the first time displayed on a portable console. However, this did not see commercial success owing to: there was no software support, no color graphics and was uncomfortable to use. This was discontinued a year later.
- **1997:** War zones were created for veterans by the researchers of Emory University and Georgia Tech where for PTSD the veterans were receiving exposure therapy. This came to be known as the Virtual Vietnam.
- **2001:** The cubic room that was first to be based on PC was the SAS Cube. This resulted in the Virtools VR Pack.
- **2007:** Street view was introduced by Google. Street view mapped the imagery and with a moving car being equipped with its patented dodecahedral camera, immersive media was the contractor identified to capture four of the five cities' imagery that was thus mapped.
- **2010:** A stereoscopic 3D mode for Street View was introduced by Google. The Oculus Rift Headset's first prototype was created by an 18-year-old entrepreneur called Palmer Luckey. A 90° field of vision had never been seen earlier and the headset featured

whereby the processing power of the computer to deliver the images was relied upon. VR saw a fresh and boosted interest due to this development.

- **2012:** A Kickstarter campaign was launched by Luckey whereby he raised $ 2.4 million for the Oculus Rift.
- **2014:** For $2 billion, the Oculus VR Company was bought by Facebook. After this VR gained rapid momentum so that it is the history of VR this was taken as a defining moment. It was announced by Sony that for the play station 4 (PS4), a VR headset, Project Morpheus was being worked upon.

A stereoscopic viewer for smartphones that was do-it-yourself and low-cost namely Cardboard was released by Google. The Samsung Gear VR was announced by Samsung which was a Samsung Galaxy Smartphone (as a viewer) using headset.

The possibilities of VR as well as the addition of innovative accessories started getting explored by more and more people. For instance, a hover board scene from Back to the Future was recreated by an independent developer-Cratesmith whereby the Oculus Rift was paired with a Wii's balance board.

- **2015:** General public could access the possibilities of VR. For instance: following the ups and downs of the Nasdaq Stock Market, a VR roller coaster was launched by the BBC. At the White House Correspondents' Association Dinner, the Oval Office' VR experience was released by the Washington Post.

A short film on the US prisons' solitary confinement, Confinement was exhibited by a media company-RYOT. The Kickstarter campaign of Glovestone turned out to be a success. The users could interact with and feel the virtual objects through these gloves.

- **2016:** By this time, VR products were being developed by hundreds of companies. Dynamic binaural audio was present in most headsets.

Using their movements and touch like with the Gloveone gloves, humans were able to interact with a computer through the Haptic interface systems that were developed. This implied that handsets were by and large operated by buttons.

The HTC VIVE Steam VR headset was released by HTC. Based on sensor tracking, this was the first headset commercially released by which within a space, users were allowed to move freely.

- **2017:** VR headsets are being developed by many companies like Google, Amazon, Samsung, HTC, Microsoft, Sony, etc. Similar to HTC's VIVE for the PS4, a location tracking tech may be developed by Sony.
- **2018:** The Half Dome, a new prototype for headset was demonstrated by Oculus at Facebook F8. Having a field of vision of 140°, this is a varifocal headset.

The tremendous progress made in VR has enabled its use in a number of ways like helping treat psychological disorders, providing immersive gaming experience, taking people who are terminally ill on virtual journeys, and teaching new skills to people. The numerous uses of VR shall be more accessible owing to the increase in the usage of smartphones.

- **2019:** This year was described by Forbes as "The Year Virtual Reality Gets Real." A large amount of momentum and interest was seen by Facebook's standalone headset, and it generated $5 million worth of content sales by selling out in a number of locations.

A shift was seen within the immersive ecosystem through the shift to standalone VR headsets from the tethered ones owing to the ease with which the average customers could use them.

For the first time the monthly-connected VR headsets on Steam crossed 1 million as per the reports on Road to VR. On April 12, with a VR kit for Nintendo Switch-Labo, Nintendo made an entry into the market of VR. The first application to sell within a year over 1 million copies was Beat Saber in the month of March.

1.3. TYPES OF VIRTUAL REALITY (VR)

3D movies, interactive video games, and television programs do not immerse the viewer/user in the virtual world either partially or fully and do not count as VR even though to create a buzzword, the marketing of these has often used this term.

Even though a VR experience that is convincing enough cannot be created by a cell phone, a search for the term 'virtual reality' is likely to give hundreds of hits on the cell phone. All the same, the definition above can be met with things computer simulations and interactive games showing that there is more than one approach that builds virtual worlds, and VR definitely has more than one flavor. Some of the bigger variations are discussed in subsections (Figure 1.2).

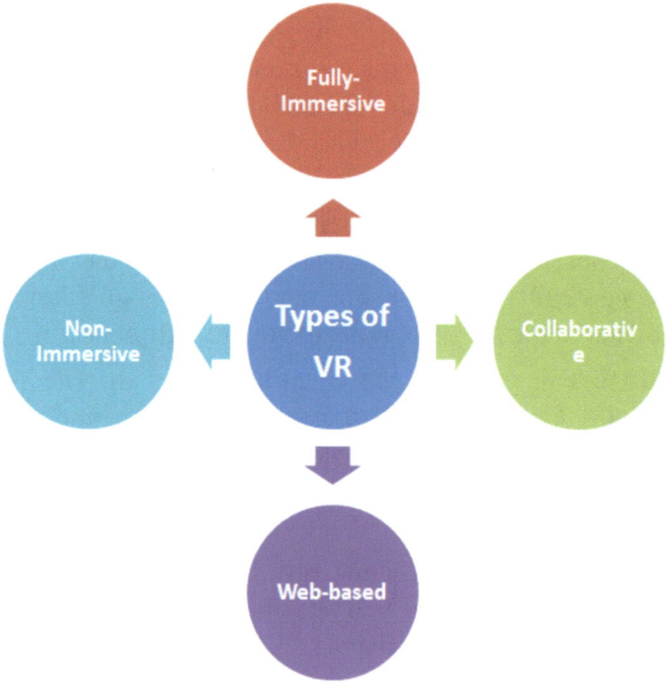

Figure 1.2: Types of virtual reality.

1.3.1. Fully Immersive

Three things are required for a full VR experience. Firstly, there should be a richly detailed and plausible virtual world that can be explored; in other words, a simulation, or a computer model. Secondly, to detect what is being done and adjust the experience in real-time, a powerful computer is required so that as fast as one moves, what is seen or heard changes at the same pace in the way it happens in the real world.

Thirdly, a computer linked hardware so that as one roams around it fully immerses one into the virtual world. Mostly, one would need to wear one or more sensory gloves and put on a head-mounted display or an HMD which has stereo sound and two screens.

On the other hand, one could move around in a surround-sound loudspeaker fitted room on which from the outside images are projected. VR equipment shall be explored in detail later.

1.3.2. Non-Immersive

If a home PC has a highly realistic flight simulator on it which uses a very wide screen, the same may qualify as non-immersive VR. Immersing in an alternative reality may not be desired or required by everyone.

A 3D model of a new building may be explored by a client where the architect wants to show it by using a mouse a desktop computer. Even though this does not fully immerse an individual, many may refer to it as a VR of some kind.

Similarly, long-lost settlements are often created in 3D by computer archeologists so that one can explore and move them around. Compared to an animated movie or some pastel drawings, a far richer experience is thus given even though the sounds, tastes, smells of prehistory are not created, and the observer is not even taken back hundreds or thousands of years.

1.3.3. Collaborative

The 5th criteria of VR, namely fully immersing the user, is not met with by "virtual games" like Minecraft and Second Life even though all the other four criteria, namely computer-created, believable, explorable, and interactive, are met by them.

At the same time, they provide collaboration which cutting-edge VR does not wherein the collaboration in a virtual world interaction is allowed with other people usually in real-time or at least somewhat close it. There is a likelihood in the future of sharing and collaboration becoming important features of VR.

1.3.4. Web-Based

One of the fastest and hottest selling technologies of the late 1980s and early 1990s was VR; however, interest in this field got over with World Wide Web's rapid rise.

A way by which virtual worlds could be built on the web (using virtual reality markup language, VRML, a technology analogous to HTML) was developed by computer scientists, but the interest of the ordinary people was drawn to the manner in which many new ways were available to them through the web for accessing real reality-new ways to shop, find, and publish information, share ideas, thoughts as well as experiences through the social media with friends.

The future of VR is likely to entail both a collaborative and Web-based approach considering the growing interest of Facebook in technology.

1.4. VIRTUAL REALITY (VR) DEVELOPMENT

Quite a bit of the research and development for projects in VR has been funded by the Department of Defense and the National Science Foundation, NASA. Sutherland received a contribution of $ 80,000 from the CIA for research purposes. The initial applications were primarily used for training exercises where they were in the category of vehicle simulator.

In case the pilot's performance was affected, the pilots were required to have a lag time (a minimum of one day) that was significant and for this, policies were framed by NASA and the military as even though the simulator provided flight experience was similar to that of real flights, it was not identical.

VR technology did not catch the attention of the public for a long time. Till the 1980s, the entire focus of the developments was on vehicle simulations. To advance the interface between human-computer (HCI) designs, in 1984, Michael McGreevy, a computer scientist started experimenting with VR technology. Due to HCI, the idea of VR was picked up by media a couple of years later, and HCI continues to play a major role in VR-related research.

In 1987 the term VR was coined by Jaron Lanier. During the 1990s, the media took to VR in all its shades. A lot of people raised unrealistic expectations from the capabilities of the technologies of VR due to the hype that was created.

The interest reduced when people realized that the degree of sophistication, they attached to VR had not yet been reached. As this happened, the term VR too began to fade away. In the current scenario, the term VR is avoided by the developers of VE, and at the same time, an attempt is made that the applications or capabilities of VE are not exaggerated.

1.4.1. How Does Virtual Reality (VR) Work?

A head-mounted display was developed that had two tiny stereoscopic screens which were positioned but a few inches right in front of the eyes leading to a breakthrough in VR. In 1989 Jaron Lanier designed the most popular VR system.

The Eye Phone, a head-mounted display, was featured in the system. For interaction and movement in the VE a DataGlove is also worn by the users. The estimated price of the system is about $205,000.

The movements of certain body parts like the hand or the head generate direct response in cyberspace, where movement is simulated through the shifting of optics in the field of vision. For instance, the scene shifts as per the turning of the head. The user gets a sensation that he/she is inside the computer created artificial world.

A set of optics with a wide-angle which can cover about 140° is used in the EyePhone, which covers most of the horizontal field of view (FOV). The image shifts when there is a movement of the user's head so that an illusion of movement is created.

Even though the virtual world stands still, the user moves. Through sensors that are embedded, the user's facial expressions too can be sensed by the glasses, and the virtual version of the user's body can be controlled through this information.

A system of glove, helmet, and a monochrome 3D was developed by a group working at NASA. Fiber optic stands running down each finger and position tracking sensors are used in a key interface device-the DataGlove so that the objects that exist within the simulated environment of the computer can be manipulated by the user.

The user actually "feels" the virtual object when the computer "senses" the virtual object being touched by the hand of the user. In the same way as an individual would handle an object by picking it up, the user is able to do the same virtually.

Robotics is one field where the applicability of the DataGlove shall be the maximum, especially when hazardous materials are being handled as also when from the safe environment of a space ship, space station, or for that matter even the earth, the astronauts need to control the repair of robots.

Simulation of vision is the primary task of VR. Creation of an immersive 3D environment through a perfect approach is the main aim of every headset. Interaction with the real world is eliminated through the screen (sometimes two screens are used where one is there for each eye) put up by every VR headset. Based on the positioning and movement of an individual the two autofocus lenses adjust (these are mostly placed between the eyes and the screen). An HDMI cable or a mobile phone that is connected to the PC help in rendering the visuals on to the screen.

There are certain prerequisites for creating a VR that is truly immersive, namely: a frame rate of minimum 60 fps; even though a FOV of 180° would be ideal, a minimum FOV of a 100° and a refresh rate that is equally competent.

The rate at which the images can be processed by the GPU per second is the frame rate, the extent to which the head and eye movement can be supported is the FOV and the pace at which the display can render images is the screen refresh rate.

Latency (the gap between the actions and the screen's response being too much) shall be experienced by the user when any of these conditions is not functional as per the requirements. To be able to trick the brain, the response needs to be less than 20 milliseconds, and this can be achieved when all the factors mentioned above are combined together in the correct proportion.

Further, unpredictability between the frame rate and refresh rate causes cybersickness or tearing and this issue too needs to be catered for. The image can get distorted when the fps of the GPU is more than the screen refresh rate. By using vertical sync (VSync), a technology that can limit the framerate to the refresh rate of the monitor, this issue can be countered.

Rift and Vive, among the various headsets available, currently have a FOV of 110°, the GearVR has 96, Google Cardboard has 90, and an FOV of up to 120° has been offered by the new Google Daydream. Further with respect to the frame rate, Oculus Rift and HTC both have 90 Hz displays, and a 60 Hz display has been offered by the PlayStation VR.

1.5. VR-NOT A REALITY BUT NEITHER A HALLUCINATION NOR DREAMING IMAGININGS

VR is neither a dreaming of imaginings nor hallucination. One lays himself/herself in full/partial or non-immersion in the simulated environment that has been created so that he/she feels that the real world is being enjoyed along with its effects by being in it through the intriguing experience imparted by VR.

Technically advanced interfaces are used to create a simulated world wherein the actual world, the senses being used (like sight, taste, smell, hearing, or touch) are controlled by the kind of experience and simulation process. In real-time, it is a paradigm shift.

In the first place the definitions of VR, simulated environment, immersion, simulation, etc., have been covered. In VR, various pioneering computer engineers, philosophers, scientists, etc., contributed their time and researched developments in this field so that it could be developed further by studying its different fields, and these have been covered here. Simulated realty's intricacies and ideas have been focused upon along with the scope for further development in the future.

VR is mainly applied in games, entertainment (films), sports, industry, space, aviation, education, medical treatments, defense (army, navy, air force), scientific research, etc., after a lot of research is done in a specific domain area by the computer engineers or scientists, innovatively, a simulated environment is created in the field where it is to be applied.

In real-time in a real-world, a replica of the actual physical objects along with their characteristics (like color, width, size, height, weight, etc.), is made through simulation. The objects are presented in a 3-dimensional feel and manner mathematically through the VE or the simulated environment.

Around 70 years ago, VR commenced its journey (the same can however, be tracked back to the 1860s even before digital technology's birth). During the late 1980s and 1990s public started noticing it.

As per the records, Jaron Lanier, the founder of VPL Research company, is credited with coining the term VR in 1987 (Microsoft acquired this company in 1999) even though many have contributed towards VR. Subsequently, a series of interface devices like head-mounted displays (HMD), glasses, and gloves were produced by the duo team of Zimmerman (the first data glove was invented by him) and Lanier.

The word experience was added to the idea of VR by Mortein Heilig, who used an oscillating fan in a 3D film so that the viewer could actually feel the force of the wind blowing on his/her face). Amongst the other pioneering scientists or engineers in the field of computers who contributed to this field are: Douglas Engelbart (contributed through the usage of a keyboard/mouse for participation), Myron Krueger (who holds the credit for CG that had audio in a video projection of the person/user in space which is similar to the modern cave automatic VE, CAVE even though with no interaction and Ivan Sutherland who holds the credit for the creation of first display that was head-mounted (Figure 1.3).

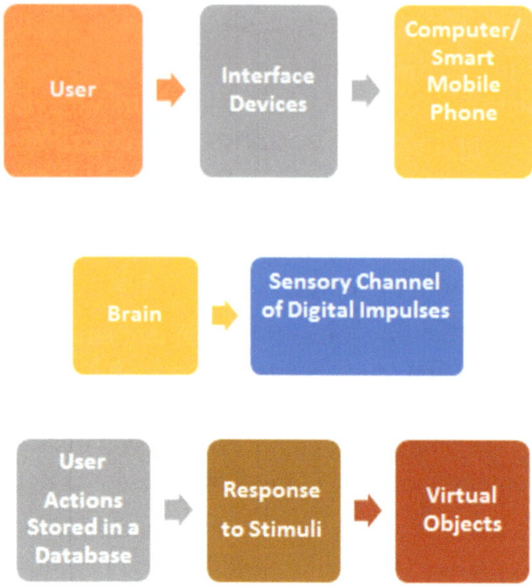

Figure 1.3: A schematic representation expressing VR.

An example of VR can be where an individual is seated on a chair that is connected with a computer interface and he/she is taken on a trip to somewhere in the galaxy like the moon or Mars and the individual enjoys the ambience of the surroundings by feeling that he/she is actually moving there.

Through VR, the experience of a participant or user gets enhanced so that in the real-world, he/she navigates and interacts with virtual objects within a virtual situation or environment. It is definitely different from being seated on a chair and enjoying a movie or theater as though the actor/actresses have been replaced by that individual.

Motion is included in a trip to the galaxy then be it the mars, moon or anywhere else so that through sight, sound, haptic, etc., the individual can move with the computer interface which does not transpire when a movie is being watched or when an individual is in a theater in a normal case scenario.

Even though the individual is not in a different setting, he/she is able to interact with the surroundings due to the motion that is generated, and in the case of a movie or a theater when the individual sit on a chair, this kind of an interaction is missing.

Whereas the interaction through a computer interface (through the creation of a virtual situation) like in the first instance is immersive (encasing the user at times), watching a movie like was done traditionally on a flat screen is not immersive.

VR is taken to be synonymous with the situation or the simulation as the two senses, sound and sight are included in this so that the participant or the user can feel that he/she is a part of that virtual world just like in the real world. A smartphone or a computer along with a head mounted display set and earphones for speakers or audio are used in this interface.

The brain which is essentially seamless can think of numerous such scenarios making the world of VR infinite which is unprecedented in the Euclidean space interfacing devices like controllers, HMDs (head-mounted display systems), gloves (other body parts as well), etc., transmit the position of the object after tracking them to the computer so that processing and storage of the same can be done further in the databases for operation.

Necessary response to the stimuli (the feedback system) is taken by the information that is stored in the databases for corresponding movement of the body parts (for instance the hand) that are moved partially or fully by the user or that of the tracking device.

In simpler terms for the effectiveness of the simulation there has to be blocking (the noise from the surroundings is blocked by the headphones so that they are not heard in the simulated environment) of the natural impulses between the nervous system and the user's body whilst the receipt of digital impulses of the virtual world is enhanced through the usage of nano-sized robots that are fused to the neurons.

Further, more research is being done as manipulation needs to be done to the central nervous system. In the normal course for small positional adjustments (bending down, turning around, repositioning, etc.), a user can use her/his own body movements whilst using a 6DOFs (a modern HMD).

To move more flexibly and incorporate more movement, like walking across two rooms, a handheld device like a space ball, joystick or a 3D mouse is used by an individual using 3 DOFs. Primarily 6° of freedom is supported by sophisticated head-mounted display units and 3° of freedom is supported by the simple ones.

Along the three-axis of x, y, and z the three translational movements are included in the first 3° of freedom (moving left or right, forward, or backward and up or down). The rotational movement (pitch, yaw, and roll) around the three-axis x, y, and z is referred to by the other three later ones (DOFs). All the same many technologies can be used to implement tracking.

There can be trackers with markers, inertial tracking as in gyroscopes or accelerators, wireless, sensors fusion, optical, tracking that is marker-less, etc., the immersive experience gets stalled if there is a latency or delay in the response created after the user's action by more than 20 ms.

A majority of times the HMD interfacing device is used in VR. Interfacing devices like Google Glass, Mobile phones, etc., are used for AR. Google's Daydream, Google's Cardboard (Daydream and Cardboard do not support positional tracking), Samsung's Gear VR, Oculus Rift, PlayStation VR and HTC Vive Pro are the headsets used.

70$ to 1500$ is the usual price range for these. The features across Cardboard to HTC Vive Pro (which is quite advanced) vary. For sensors and their uses, positional tracking, and the latency time, proper research has to be done on the VR headsets specifications with respect to these features.

1.6. PROS AND CONS OF VIRTUAL REALITY (VR)

Every technology has its advantages as well as disadvantages, and VR is no different. Simple question can help understand this point. For instance, how many people would prefer to be operated by a surgeon who has only read books or watched someone else operate rather than a surgeon who has been trained on VR?

Similarly, before taking an actual driving lesson with a physical car in real-time, many people would first prefer to practice on a car simulator. Many people would be comfortable and at ease flying with a pilot who has experienced flying and landing in a simulator with the same airport set up as compared to one who has never done so.

A point that has been brought forth by the critics of VR is that people may get so lured by the alternate realities that they may start neglecting their lives in the real world however, almost everything then be it the internet, computer games, radio or TV have faced a similar criticism.

At a certain point of time, ethical and philosophical questions get involved for instance: what actually is real? Further, no one else can real say or decide as to what is the best way to spend one's time. One need not use VR if he/she does not want to, and at the end, just like many other technologies VR actually takes away nothing from the world that actually exists around us in other words, the real world.

For at least the last 25 years, VR has promised a lot many things in the computing world, but the same have not been fulfilled. The mainstream is not using VR in the manner smartphones, computers, or even the internet is being used, whereas VR technology is being heavily relied upon by the military, medicine, science, and architectural fields.

In 2014 Facebook acquired Oculus, a VR company whereby renewed interest was shown in this field, and it seemed for some time that everything would finally change. Basically, using the web and the internet Facebook allows people to share various things with their friends or even the public at large.

Facebook is now thinking of exploring numerous social possibilities in collaboration with VR where for instance people can attend functions like weddings, parties, get-togethers, etc., remotely, and live them time and again virtually; recording historical events so that they can be re-lived time and again for perpetuity. Even if these possibilities do not actualize in the immediate times, there is definitely a great deal of scope for VR.

Sadly though, the interest and great deal of optimism that accompanied the arrival of Oculus has all but fizzled out. For instance, in 2019, both BBC and Google stopped all their major VR projects. Even though the VR technology sounds quite transformative and persuasive considering that several million VR headsets have been sold by Sony, it needs to be borne in mind that billions of cell phones have been sold around the same period showing that this technology is still far from being that popular.

At least for now, this field shall continue to garner only a niche interest and shall increasingly be continued to be used through its imaginative applications in the field of medicine, science, and architecture amongst the others.

1.6.1. Advantages

In a number of cases, post-traumatic stress disorder and phobias like the fear of heights, spiders, flying, etc., have been treated by the usage of VR. In the academic field, too, this therapy has been quite effective where patients are being offered VR-based applications by many commercial settings. Even though live sessions for training have been widely used for standardized patients a number of advantages are offered by simulations that are computer-based.

The main aim of these simulations was the reduction of psychological distress and enhancing decision making as well as performance in real health emergency by increasing exposure to life-like emergency situations.

1.6.2. Disadvantages

As per some psychologists, the user could get affected psychologically by getting immersed in the make-belief VEs. They fear that where the user is placed in violent situations by the system and more so as the perpetrator of the violence, it could lead to desensitization of the user.

There is thus a fear that a generation of psychopaths could be the result due to the VE entertainment systems. Furthermore, it could become addictive to engage in VEs. Criminal acts are another concern that has been emerging. It is difficult to define crimes related to sex or murder in the virtual world.

One begins to wonder whether authorities can punish a crime committed in the VE and if so, at what stage. Stimuli that are generated within an environment that is virtual has been seen through studies to have result in real emotional and physical reactions whereby it could mean that real traumatic emotions can be felt by the victims of a virtual attack.

1.6.3. Challenges

Decreasing the time taken for the building of VR, finding natural ways by which the user can interact within an environment that is virtual and development of tracking systems that are better see to have been the major challenges in the world of VR since the very early stages of VR.

Furthermore, input devices that are specifically meant for the VR applications are not being worked upon by many companies. Most of the times, technology meant for other disciplines has to be relied upon and adopted by the VR developers in the hope that the company that is producing these does not shut down.

Also, as it takes longer to make a realistic environment (where it is directly proportional to it), it may take a long time to make a VE that is truly convincing. In order to accurately duplicate a real room in virtual space, even more than a year may be required by a team of programmers.

1.7. CONCLUSION

After reading the whole chapter, it can be said that VR basically is the use or application of computer technology for the creation of simulated environment. Different from the traditional user interfaces (UI), VR places the user inside a simulated experience. Users are immersed and able to interact with 3D worlds instead of just seeing a screen in front.

However, the technology of VR is a field which is still evolving as well as diversifying. Its evolution has almost undergone 20 years of research and development, and now has reached, on the one hand, a decent maturity, while on the other hand, it has still not reached to the mainstream in the professional as well as home or entertainment world.

In total, VR develops an immersive artificial world that can appear quite real by the application of computer technology. Through VR, users/viewers can look up, down or in any direction, as if they are actually present there. VR has various use-cases, such as entertainment and gaming, or acting as a sale, educational, or training tool.

REFERENCES

1. Barnard, D., (2019). *History of VR-Timeline of Events and Tech Development*. [Online] Virtualspeech.com. Available at: https://virtualspeech.com/blog/history-of-vr (accessed on 29 April 2021).
2. Franchi, (1996). *Virtual Reality: An Overview.* ERIC digest. [Online] Ericdigests.org. Available at: https://www.ericdigests.org/1996-2/virtual.html (accessed on 29 April 2021).
3. Mandal, S., (2013). Brief introduction of virtual reality and its challenges. [eBook] *International Journal of Scientific and Engineering Research.* Available at: https://www.ijser.org/researchpaper/Brief-Introduction-of-Virtual-Reality-its-Challenges.pdf (accessed on 29 April 2021).
4. ResearchGate, (2020). *(PDF) Virtual Reality: An Overview.* [Online] Available at: https://www.researchgate.net/publication/343095838_Virtual_Reality_An_Overview (accessed on 29 April 2021).
5. Strickland, J., (n.d.). *How Virtual Reality Works.* [Online] HowStuffWorks. Available at: https://electronics.howstuffworks.com/gadgets/other-gadgets/virtual-reality.htm#pt9 (accessed on 29 April 2021).

CHAPTER 2

Virtual Reality Systems

CONTENTS

Virtual reality (VR) systems have been discussed in this chapter by considering different types of VR systems and on the basis of immersion and degree of presence. In the first few sections, VR system types and components have been elaborated by understanding their working. Apart from that, tools and software such as 3D modeling software and 2D graphics software have been considered to discuss the working of the VR systems more efficiently.

The VR interfaces that have been considered in this chapter helps in understanding different paradigms of VR used in current times. At the end of the chapter, the VR tracking systems has been described in which different applications of VR and the use of tracking systems in it has been discussed. At the end, a comparison is made for immersive and non-immersive VR user interface is done. This chapter will allow a reader to understand the functionality of VR systems in an efficient manner.

2.1. INTRODUCTION

The ranking in the three main categories of virtual reality (VR) is done as per the degree of presence or the sense of immersion provided by it even though categorizing all the VR systems is quite difficult. The strength or the power with which the user's attention is focused on the task that is being performed can be referred to as the presence or immersion.

Several parameters have been believed to be responsible for the immersion presence, which includes the level of interactivity, stereoscopic view, update rate of display, complexity of the image and field of regard.

For instance, the users' experience of the sense of immersion shall get enhanced when instead of a microscopic view a stereoscopic view is provided of the virtual environment (VE). It must be always borne in mind that only through a complex interaction of the various factors that are involved the level of immersion is reached and no single parameter can bring about that result in isolation.

2.2. TYPES OF VR SYSTEMS AND HARDWARE

As per the usage of the technological supply, the classification of the various types of VR systems is done. Various interaction and hardware devices that are displayed represent the various equipment and supplies. Ranging from semi-immersive or desktop; the fully immersive VR to the augmented reality (AR), the level of immersion provided by the VR systems lays down the basis for their classification. The explanations that follow show the different types of VR systems that perform varying functions and use varied technological devices.

2.2.1. Immersion Systems (Fully-Immersive)

VR systems that are fully immersive provide the most direct experience of VR. In these systems which are in all likelihood used the maximum, the user either uses a head-coupled display of some kind like the BOOM or the binocular omni-orientation monitor or wears an HMD (Bolas, 1994).

2.2.2. Head-Mounted Displays (HMDs)

In front of each eye, a small monitor is placed in an HMD through which stereo, monocular or bi-ocular images can be provided. Each eye is presented with an image that is slightly different in a manner that is similar to the shutter glasses whereby stereo images are provided.

Due to the HMD optical system, the images on which the user is focused shall be quite far away, and herein lays the major difference as the two screens are placed quite close to the eye (50–70 mm). By displaying images on a single screen, a monocular image can be provided, and with a display of identical images on each screen, a bi-ocular image can be provided.

The resolution of the image can be increased by the use of the cathode ray tubes (CRT) in the expensive HMDs though the small liquid crystal display (LCD) panels are more commonly used. The view of the real world may be fully or partially obstructed for the user by the HMD design, and the computer-generated world's field of view (FOV) is enhanced.

If the user's moves their head in any direction, a visual image shall be received by them, which basically means that the user is provided a 360° field and herein lays the advantage of this method.

None of the approaches discussed earlier can equal the sense of presence given by the all fully immersive systems. However, various parameters come to play in providing the sense of immersion like the HMD's FOV, the update rate, the resolution as well as the illumination and contrast of the display.

So that the level of realism achieved is satisfactory, a high level of technology and computing power with the involvement of a great cost is required by the fully immersive VR systems and constant development is taking place in this field so that the technologies involved can be improved. The main field of development and research are reduction in the weight and size of HMDs, FOV vs. resolution trade-offs and the reductions in the system lag times.

The user is required to wear a data glove and HMD through which the head movements of the user's head can be tracked in the VR system that is of the immersion kind and then the view is changed. An example of a fully immersion technology is the CAVE. CAVE uses large projection screens and is designed in such a way that implementing it can help to deal with as well as treat the challenges creating a visualization tool that is one-to-many.

The visual and audio perception of the user is encased by this kind of a VR system in the virtual world, and to make the experience fully immersive, all the outside information is cut off. Even as this technology is expensive, it has certain disadvantages as well, like the images being less determined, simulator related environmental problems and burden.

The user gets a feeling of completely being a part of the VE when full immersion of VR technology is used. One of the applications of VR technology where this can be taken as an illustration is a virtual walk-through of buildings.

The figure (add the image of simulator) shows some examples where full immersion of VR is used. The "Light Vehicle Simulator" of fully immersed has been represented, and this is used for training requirements. This simulator gives an experience to the users who drive vehicles at the mining sites whereby they learn to deal with and respond to various kinds of risks and emergencies (Figure 2.1).

Figure 2.1: Light vehicle simulator.

Source: Image by Sidney Daily News.

2.2.3. Non-Immersive System

The desktop VR or the non-immersive system does not use any input devices and is sometimes called Window on World (WoW) systems wherein without any additional devices like the HMD, it is a window to the virtual world based on the displayed screens.

The implementation of VR techniques that is the least immersive is the non-immersive systems. Using a standard monitor of high resolution, through a window or a portal the VE is viewed using a desktop system.

Conventional means can be used for interaction with the VE like for example, the keyboard, trackballs, and mice or 3D interaction devices like DataGLove or SpaceBallä may be used to enhance the experience.

Graphics performance of the highest level is not required for a non-immersive system nor is a special hardware necessary which is its biggest advantage. It can be implemented on PC clones of high specification.

As a result, for many applications, these systems are the most cost-effective VR solutions. At the same time, it also implies that due to the low cost involved, more sophisticated implementations shall constantly keep outperforming it, the systems shall provide nearly no immersion sense, and the current 2D interaction devices shall limit it to a certain extent.

Moreover, where the important factor is perception of scale, these systems are not of much use. All the same in the near future, for VR, the popularity of these kinds of systems is likely to increase. This is mainly due to the reason that for the transfer of virtual worlds and 3D model data through the internet, the de-facto standard likely to be adopted is the virtual reality modeling reality language (VRML).

Many of the proprietary VR authoring tools do not work well on a PC, whereas the software of VRML works quite well for the user of a PC desktop. Moreover, the commercial possibilities of desktop VRML are being explored in general by many suppliers of commercial VR software whereby VR capability is being incorporated into their software.

Virtual Reality Modeling Language

A programming language for the creation of virtual worlds. Using a VRML viewer, you can take a virtual tour of a 3D model building, or manipulate animations of 3D objects. Hyperlinks to other sites and files can be embedded in the world you visit.

The desktop system is the most widely used VR system wherein the virtual world is displayed through a standard computer monitor. Graphic quality of satisfactory levels, convenience, and comfort of the user and costs that are lower can be achieved through these systems even their interaction and presence is relatively low.

Amongst the VR systems, the cost in the desktop VR system is the lowest, and the immersion in this is the least. Non-immersive kind of VR is used mainly in education and has the least complicated of the components.

The virtual world is another form of the desktop VR system. In the field of education, it helps the user to observe the information while providing enhanced understanding and learning to the user. Through various avatars, interactions among the humans can be carried out through the virtual world's pull together systems.

For the construction of the virtual worlds, many open-source software packages are available like Active World, Open Croquet, Second Life and Open Simulator.

2.2.4. Semi-Immersive System

Hybrid systems are the third kind of VR systems. Being a development desktop VR semi-immersive includes additional devices like the Data Gloves. Using physical models and having a high level of immersion the desktop VR system's simplicity is maintained by it.

The real environment that is recognized is set up with the VE that is displayed in semi-immersive. Tracking sensors, displaying, and user interfaces (UI) are the requirements for building a semi-immersive system.

A graphics computing system that has relatively high performance is comprises the semi-immersive system and this is coupled with either of the following:

- Multiple television projection systems;
- A large screen projector system;
- A large screen monitors.

These projection systems draw similarities to the IMAX theaters in many ways. The user experiences an enhanced feeling of presence or immersion in these systems due to the usage of a wide FOV.

All the same, an important consideration is the quality of the image that is projected. The quality of the colors, textures, ability of the user to read text on-screen, and the ability to define shapes is determined by the resolution and to avoid distortions, it is important that the projected image's geometry is calibrated to the screen's shape.

Multiple projection systems may be required to achieve the highest levels of resolution, which are far more expensive compared to the projection systems that provide a resolution of 1000–3000 lines.

Compared to the non-immersive systems a greater appreciation of scale and sense of presence is provided by the semi-immersive systems. Furthermore, as compared to HMDs images with a far great resolution can be provided which enables the sharing of the virtual experience.

Educational applications can be majorly benefitted by this as the head-mounted immersive systems do not allow simultaneous experience of the VE. Moreover, when the graphics system is synchronized with some kind of shuttered glasses and they are used, stereographic images can also be achieved.

By embodying the reality scene with objects of computer graphic real-world and VR attributes are incorporated in the semi-immersive system. The users enter and control the input of this kind of a system through the usage of a keyboard, glasses, interaction styles, joystick, and mouse. The user can at times wear DataGLove or glasses and use hands to interact in this.

To enable interaction of the user with the real environment, the displayed information like graphs, texts, and images jut onto the transparent screen.

2.2.4.1. Shutter Glasses

When semi-immersive systems are considered, an important technology is the liquid crystal shutter (LCS) glasses where over each eye a liquid crystal lens is placed within the lightweight headset that is used.

Slightly different images of the same scene must be regarded by each eye of the observer so that in a scene depth can be perceived and on stereopsis works on this principle. In real life, the scene is viewed from slightly different positions as on the head, the eyes are placed slightly apart.

On the display system slightly varying views of the left and right or the stereo pair of the VE is displayed sequentially through the graphics computer that is used. The image that is projected or produced on the VDU are either blocked or passed by the glasses so that the stereoscopic effect can be produced.

The left eye lens is switched on when the left image is displayed so that the screen can be seen by the left eye of the viewer, whereas the view of the right eye is blocked with the right eye lens being off. The opposite transpires when the image displayed is the right one. The user is unable to detect this switching of the images as it occurs very rapidly, and one constant 3D image is seen as the two images get fused in the user's brain.

Some of the examples of this product that are available include the 3D max Shutter Glasses System and the Crystal Eyes Shutter Glasses.

However, high costs are again involved in the implementation of this VR if high performance is to be achieved. When compared to a desktop system, the setting up of a projection screen is not only far more expensive but is also more difficult.

To add on, with these devices, the current interaction devices have problems. The system may be used for certain applications, and in the first place, these need to be considered. The use of an interceptor or joystick may be possible in a simulation system, and the same can be interpreted as the input for the flight control by the aircraft model.

As for no other application is the simulator used, it is acceptable; however, there may be multifarious uses to the semi-immersive installation, and then it can become a problem as many interaction strategies of different kinds may be required.

One of the main advantages of the semi-immersive system is the multi-users aspect, and in the second place, multi-user issues need to be considered. With the developments in this technology, one of the issues that need to be considered is the handing over of control between the users.

2.3. COMPONENTS OF VR SYSTEMS

An overview of the VR systems is shown in this section which takes place from the hardware to the software to human perception and understanding the working of the VR systems in its entirety. Two major subsystems, namely the software and hardware comprise the VR system.

The input/output devices and the VR engine or the computer further make up the hardware and the software can be divided into data base and application software in the manner illustrated in Figure 2.2.

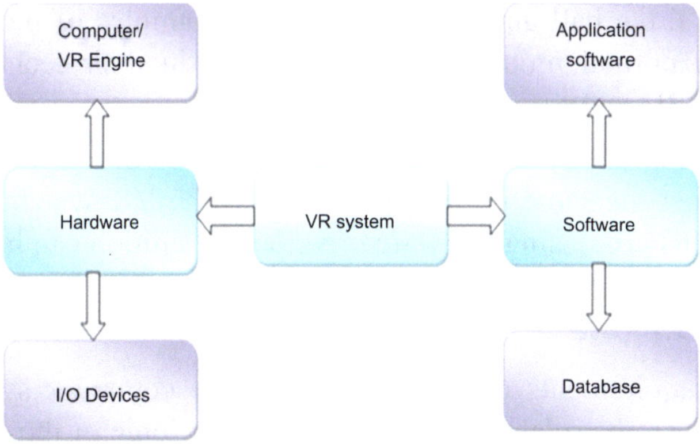

Figure 2.2: Components of a VR system.

Source: Image by Atlantis Press.

2.3.1. Virtual Reality (VR) System Hardware

The computer system or the VR engine, the output and input devices form the major part of the hardware, as shown in Figure 2.3.

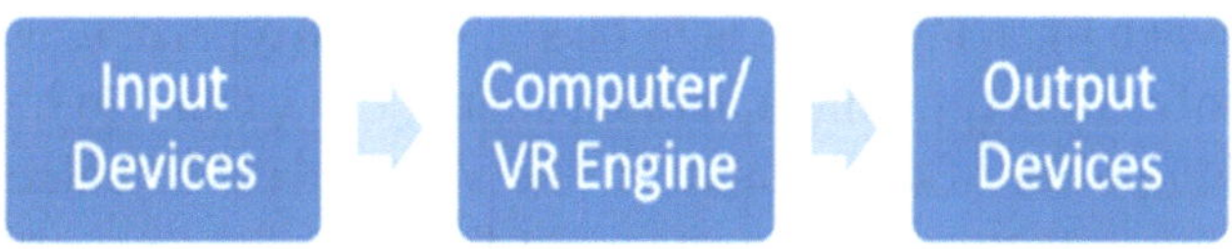

Figure 2.3: Major parts of VR system hardware.

2.3.1.1. Input Devices

The user's interactions with the virtual world are made possible through the input devices. The action of the user is conveyed to the system through signals sent by them so that the reactions sent back to the user via the output devices are appropriate and in real-time. Tracking device, voice device, point input device and bio-controllers are its main classifications.

The position of the user can be tracked through the position sensors, also known as the tracking devices and include ultrasonic, gyroscopic, and mechanical sensors, bio, or muscular and neural controllers, electromagnetic, optical, and data gloves.

Space or force ball and 6dOF mouse are examples of the point-input devices. The normal mouse has been adapted in their technology with capability for 3D and extended functions.

Humans commonly interact through voice communication making it but natural that the same be incorporated into the VR system as well. To accomplish this processing software or voice recognition can be used.

2.3.1.2. VR Engine

As per the requirements of the application, the computer system or the VR engine has to be selected in the VR systems. Some of the VR system's most time-consuming tasks and important factors are image generation and graphic display.

The application field, graphic output that is required as it is responsible for generating and calculating the graphic output that is required, level of immersion, the input/output devices, object rendering, mapping, simulation, lighting, texturing, and display that is in real-time help in deciding or making the choice of the VE engine. The computer also serves as an interface with the input/output devices and handles the interaction with the users.

The processing power of the computer is the major factor to be considered when the VR engine is to be selected. In a particular time frame, the number of senses (graphical, haptic, sound, etc.), that can be rendered is the processing power of the computer. The VE needs to be calculated by the VR engine every 33 ms and more than 24 fps in real time simulation needs to be produced.

At the same time, stereoscopic vision should be produced by the associated graphic engine, and it must have the capability of doing so. High speed communication network should be there to interconnect distributed computer systems or graphic accelerator that should be powerful where the VR engine could be a standard PC that has a higher processing power.

2.3.1.3. Output Devices

In order to stimulate the senses, the VR engine gives feedback to the output devices, and they in turn, through the output devices that correspond to it, pass it on to the users. Based on the senses, the output devices can be classifying into: audio (aural), smell, graphics (visual), taste, and haptic (force or contact). Taste and smell are still not as common in the VR systems as audio, graphics, and haptic.

Immersion of a higher level is provided through the two options of HMD and stereo display monitor which, for graphics, are the most common options. A 3D view of the virtual world is provided in the HMD by the two views that are independently produced and then interpreted by the brain.

In VR the sound or audio channel is quite important whereby only visual is more important than this. VR application is made even more realistic by producing different sounds from varied directions to produce a 3D sound. User is able to feel the virtual objects through the haptic. Mechanical devices and electronic signals are used to achieve this.

2.3.2. Hardware

Based on human motions, the senses of the user are overridden through the stimuli produced by the hardware. Sensors are used by the hardware to track the user's motions like button presses; eye and other body part movements, and controller movements.

Complete VR system does not constitute of just the software and hardware so that the physical surrounding world too is considered. Of equal importance is the interaction of the user with the hardware and the user himself/herself.

Sensors in the VR hardware act as transducer so that the energy received by from the electrical circuit it is converted into a signal. For conversion, the energy is collected by the receptor in the sensor, and there are sense organs like the ears and eyes in the user for a similar purpose. Every movement of the user in the physical world has a specific configuration space, and these are configured or transformed accordingly.

2.3.3. VR Devices

For VR technology to work, certain hardware products are used in the VR devices. VR system's main components are discussed hereinafter. The high-level view of virtual world generator (VWG). For VR experience's appropriate view is displayed based on the inputs that are received from the surroundings of the user and the user himself/herself.

2.3.3.1. Personal Computer (PC)/Console/Smartphone

Inputs and outputs are processed sequentially through computers. There is a requirement of a large amount of computing power for the content to be created and produced so that an important part of the VR systems is then the

PC, the smart phone or the consoles that are used. What the user perceives and views inside forms the VR content making it as important as any other hardware.

2.3.3.2. Input Devices

The manner in which a user shall communicate with the computer is determined by the input devices, which also provide a sense of immersion to the user. To make the VR environment as natural and intuitive as is possible, users are helped by the input devices to determine their way and gain a sense of immersion.

However, as of now, the technology that is available does not support this as it is not that advanced. Tracking balls/force balls, data gloves, on-device control buttons, bodysuits, motion platforms (virtual omni), joysticks, controller wands, trackpads, motion trackers and treadmills are the input devices most commonly used.

2.3.3.3. Output Devices

The sense organs are stimulated by the output devices. The environment or the VR content is presented to the user through the output devices, and this is the device that generates the feeling of immersion. The perfect ideal manners of the human senses cannot be stimulated by the existing state-of-art VR system and the output devices as well that are currently in use are underdeveloped. Auditory, haptic, or visual displays comprise the output devices.

2.3.4. Software

The software incorporated is as important as the output-input hardware and its coordination. It is the software that analyzes the incoming data, manages the input-output devices, and generates appropriate feedback.

The software must be able to manage the entire application, which is essentially time-critical. For instance, to sustain the feeling of immersion, not only the input data needs to be handled timely but also receipt of system's response should be sent to the output displays promptly.

The developers can from the scratch build their own VWG by starting with a kit for software development that is basic (SDK) taken from a vendor of VR headset. An interface that can be used to call graphical rendering libraries and track data as well as the basic drivers are usually provided in a software development kit (SDK). For specific VR experiences certain ready-made VWG exists, and they also have an option whereby high-level scripts can be added.

2.3.5. Audio

VR audio is an important part of the whole set up so that immersion can be achieved by the users and their senses can be stimulated even though these audios may be simpler technology wise when seen vis-a-vis the visual components.

In conjunction with a headset, the users are often given an option to use their own headphones by most of the headsets used for VR. Their own headsets that are integrated may be further included in other headsets.

An illusion of a 3-dimensional world is given through multi-speaker, positional audio (often referred to as the Positional Audio) via which VR audio works. Cues are provided so that users' attention can be gained through positional audio as visually, some information may not be provided to them so in a manner it helps to see through the ears. The surround sound systems in home theater uses this technology, and it is fairly common.

2.3.6. Human Perception

For gaining of maximum human perception and no side effects, it is essential that optical illusions and physiology of the human body is understood. Various stimulus, sense organs and receptors are used by the human senses.

The realReal-world is simulated in VR, making it important to know as to how the senses of the user can be fooled as well as the quality of subjective viewing that is required and the important stimuli. Various senses like touch and hearing, among the others, are provided with information once the visions seen by the humans are passed on to the brain. For the VR system to function well, it is essential to system synchronize all the stimuli with the actions of the user.

2.4. VR SOFTWARE AND TOOLS

The software and tools that are required to use and develop VR systems are available across the globe but are still growing as per the increasing usage of VR technology in numerous fields across the world. Recognition of the software for those applications is the key factor for realizing the applications of the VR system.

Two main kinds of VR software are available currently, namely the authoring systems and the toolkits. Whereas complete development environments are some of the VR applications, others are the framework. Many aspects go into the construction of a VR environment, and there needs to be an integration of these into a single package.

Using toolkits in a library, a proficient programer can introduce a set of functions and create a VR application. All the same, the authoring systems have graphical interfaces but do not have resource to detailed programming and are essentially a simple program. As per Onyesolu, there are four main components of VR technology's software, namely: 2D graphics software, 3D modeling software, VR simulation software, and software to edit digital sound.

2.4.1. 3D Modeling Software

In VR environments, geometry objects are built using a computer through this program that inspires 3D images. This component of software uses tools like:

- ***Autodesk 3D Max:*** 3D Studio MAX is another name for this. This is quite extensive and has many applications for 3D that can be used in architecture, films, and video games. Apple Macintosh operating system and Windows are used to work this.
- **GL Studio:** 3D graphics that are interactive with the user interface are made through this. To create virtual worlds with interfaces the OpenGL source code and C++ programming language are the basis for GL Studio.
- ***Electric Image Animation System (EIAS):*** Movies like Hollywood used this extensively, and this application is mainly used to generate films. Animated 3D environments are created through this.
- ***Maya:*** With a 3D virtual world this finds applicability for creation in the gaming industry, on TV and creation of movies. It has compatibility with a number of operating systems like Linux, Mac, and Windows, Cobalt, AC3D, massive, and Cinema 4D. Either the drawing program or 2D graphics editor can be called

by it. For providing support to the visual details, it is used in 3D constructions to operate objects and play.

2.4.2. 2D Graphics Software

Through the usage of graphics tablet, mouse, or hardware similar to it images and drawings are created as well as manipulated through 2D graphics software. Drawings to find use for 2D software like for electronic diagrams, electrical, fonts in a computer and topographic maps. The components of VR are integrated through this program with the manner in which the rules are set, objects conducted so as to guide the VR environment that follows this.

2.4.3. VR Simulation Software

OpenSimulator or OpenSim is an example of this kind of simulation software. A 3D environment is created is created through this 3D application server and may prove beneficial for developers so that they can build and improve various applications like chat application among avatars to develop the application, series of programming languages are supported by OpenSim like C#, Jscript, and Linder Scripting Language.

Ogoglio is one other example of VR simulation. For artistic collaboration, online spaces can be built through this open-source 3D graphical stage. As a scripting language, it is similar to JavaScript. It can run over any browser being compatible with Linux, Windows, and Solaris operating system.

FlexSim DS is another example. Distributed VR environments are built through networks using FlexiSim DS and is considered one of the advanced 3D simulations.

2.4.4. Digital Sound Editing Software

Sounds are edited and mixed in VR environments through editing software of this kind with other objects that are available in the same environment. Some of the sound editing software includes: Audiobook Cutter Free Edition, FlexiMusic Wave Editor, Media Digitalizer, Creative Wavestudio, mp3DirectCut and Goldwave.

By integrating the virtual reality modeling language (VRML), which is the VR language with a programming language, 3D objects are built and constructed in the VR environment with UI and animation. Scalable vector graphics (SVG) can be represented through VRML which is an easily written modeling language.

An additional plug-in can be used to view VRML on a Web browser. Alternatively, it can even be used alone for viewing on a viewer or player. The inception of VRML language was in 1994, and since then it has improved via various versions whereby this language is used to create and construct most of the interactive virtual world and 3D environments.

Currently, for the construction of VR, the 3D language that is most commonly used is VRML. So that the structure data can be easily accessed, most of the times, VRML is used according to feature. Independently, on a computer, VRML files can be executed and downloaded. All the same to run and operate the same on a viewer or player, an additional plug-in is required.

2.5. VIRTUAL REALITY (VR) INTERFACES

2.5.1. Data Gloves

Commands can be simply gestured to the computer through the use of data gloves. Whilst operating the BOOM or wearing a head-mounted display, it may get quite tricky to type commands on a keyboard rather they would be more like punches, and the data gloves help to change modes on the computer (which has been programmed accordingly) through simple gestures.

Pointing downwards may mean zoom out and zoom in could be denoted by pointing upwards. The computer may end the program by a signal such as shaking of the fist. Sometimes, the computer is programmed in a manner so that the hand movements are mimicked in the simulation like for instance, when they conduct a virtual symphony, their own hands can be seen.

2.5.2. Wands

The interface devices of the simplest kinds are wands which come in different variations and sizes. In order to display data or control the variables in a simulation, on-off buttons are incorporated into most of them. As per the application, their manner of response and design is tailored.

Six degrees of operational freedom is incorporated into most wands whereby an object's position can be changed and oriented in any direction namely: upwards or downwards, forward, or backward, and right or left by simply pointing the wand towards an object.

2.5.3. Stair Steppers

The manifestations of interface devices are limitless, and stair steppers are an example of this. A stair stepper was outfitted as part of a battlefield terrain that had been stimulated by engineers in an army research lab where the device had sensing abilities to detect the intensity, speed, and direction of the movements made by a soldier in response to the scenes of a battlefield that were projected onto a display that was head-mounted. By changing the difficulty level of the steps that were to be climbed, feedback was provided by the stair stepper.

2.6. VIRTUAL REALITY (VR) SYSTEMS (FIGURE 2.4)

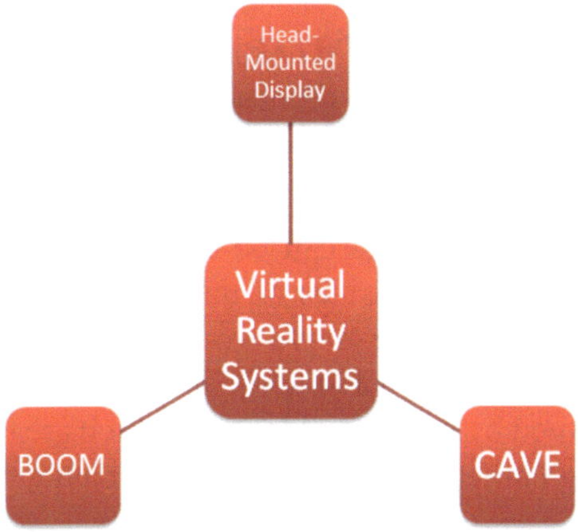

Figure 2.4: Major virtual reality systems.

2.6.1. Head-Mounted Display

The flat images are provided with depth through the head mounted displays (HMD) that look like motorcycle helmets which are oversized and are really viewing screens that are portable. A viewing screen can be seen through the two lenses when one looks inside the helmet.

Two images that are slightly different from each other are projected by the computer when simulation begins, where one shows the image as it would appear when viewed through the left eye and the other as would be viewed through the right one.

The brain then fuses these two stereo images into one image that is a 3D one. In relation to a tracking device that is stationary, the head movements are signaled through a device on top of the helmet so that movements can be tracked. The computer updates the simulation continuously so that the new perspective is reflected whenever the users move their head backwards, forwards, sideways or when they merely look in another direction.

VR operators favor head-mounted displays as the entire surrounding environment is blocked by these devices where it is desired that the wearer can fully absorb the VE like for instance, in a flight simulator. The entertainment industry makes rampant use of these displays. The most commonly used interface devices along with head-mounted devices are wands and Data gloves.

2.6.2. BOOM

Similar to a head-mount, the BOOM or Binocular Omni Orientation Monitor does not have the fuss of a helmet. From a rotating, two-part arm the viewing box of BOOM is suspended. One enters the virtual world by something as simple as placing the forehead against the two eyeglasses of BOOM.

The perspective of an image can be changed by simply grabbing the handles that are placed on the sides of the viewing box and just the way one would have done is real life, move around the image. To look at it from below, it needs to be bent down and to see it from behind, just walk around. The interface is served by the control buttons which are there on the BOOM handles even though other interface devices like data gloves can be hooked onto.

2.6.3. CAVE

CAVE is kind of a VE that is the most "immersive." On the floor and walls of a room-sized cube stereo images are projected by it to provide illusions of immersion. The CAVE can be walked into freely by a number of people at the same time who are wearing lightweight stereo glasses.

Interaction with virtual images and navigation through the VE is made possible for the user through data gloves, hand held wands and joysticks, among other input devices. To create more "corporeal" interfaces and make the immersive experience richer, employment is made of directional sound, voice recognition, tactile, and force feedback devices and other technologies.

Through the respective usage of a BOOM device, a CAVE system, and a head-mounted display three networked users can meet in the same virtual world even though they are located at different places. The same VE is seen by all three of them but from their individual point of view.

Each participant sees the other as a virtual human. As a team, all three individuals can pair up as a team where they can communicate and see each other in the virtual world. The chart below shows the level of immersion as measured.

2.7. VIRTUAL REALITY (VR) TRACKING SYSTEMS

For any VR system, the intrinsic components are the tracking devices. The processing unit of the system is communicated with by these devices so that the orientation of a user's point of view is informed to it. The speed and direction of movement of the user is detected by the tracker when the user is allowed to move freely within a physical space by the system.

VR systems use various kinds of tracking systems. Six degrees of freedom (DoF) can be detected by them, which include the position of the object within the three coordinates of space-x, y, and z as well as the orientation of the object. The yaw, roll, and pitch of the object are included in the orientation.

This basically means that when an HMD is worn by a user, as the user looks left, right, up, or down, the view shifts. Even when the angle of the gaze is not changed, it changes when the head is moved backward or forward and also when it is tilted at an angle.

Right images are sent by the CPU to the HMD screens after the CPU is informed by the trackers on the HMD as to where an individual is looking.

A device is present in all the tracking systems through which a signal is generated, then there is a sensor through which the signal is detected, and finally a control unit through which information is sent to the CPU. The sensor component is required to be attached to the user or the equipment of the user in some systems.

Signal emitters are placed at fixed points in the environment in this kind of a system. In certain other systems, it works the other way around where the sensors attached to the environment surround the user and emitters are worn by the user.

Emitters send signals to the sensors in many forms like electromagnetic signals, mechanical signals, optic signals, and acoustic signals. There are disadvantages and advantages inherent in all kinds of technology.

2.8. CRITICAL ASPECTS OF VISUALLY COUPLED SYSTEMS (VCSS)

Immersive VE systems or the VR systems of all kinds have a visually coupled system (VCS) which forms its heart. Sensors, space trackers, display generators and helmet-mounted displays comprise VCS.

VCS, its components, performance requirements and interfaces are described in this chapter. Selection that has been done very carefully, along with matching of a series of technologies from the point of software integration and hardware interface are required for VCS components' integration. Furthermore, the task requirements need to be examined so that it is ensured that their consideration has been done in the design.

2.8.1. Visually Coupled Systems (VCSs)

A head tracking system, a sensor system that is steerable and a computer graphics (CG) system comprise a generic VCS.

Head Tracker: Relative to a fixed position in space, the orientation (elevation, roll, and azimuth) and the position (jc, y, and z) are determined by the head tracker.

Helmet-mounted display: A binocular or a binocular and monocular image of the scene/objects in the user's line of vision is provided to the user through the helmet-mounted display. The focus of the image may be at a point in the near field, or it may be collimated at infinity depending upon the application. Either the steerable sensor system or the CG system conveys the display information.

When the steerable sensor or the CG system conveys information to the helmet-mounted system visuals are computed through the usage of headline of sight data to produce a VCS.

2.8.2. Head Tracker Issues

Very little importance is given to the parameters that have an effect on the performance of the system, and more often than not, the head tracking system is taken for granted. There shall be unacceptable lags in the images that are presented to the user when the integration of the head tracker with the VCS is not done carefully. The contribution of the head tracker towards the overall lag needs to be understood even though within a VCS, it is not the sole source of lag.

At a certain point in time, the only parameter considered to be critical in a space tracking system was the time update rate. A number of parameters need to be considered even though even though an explanation for tracker performance can be given through phase lag performance where phase lag is a newly introduced term. Dynamic accuracy, static accuracy, update rate and latency are the other parameters.

The consequence that each of these parameters, the interdependencies that exist between them, and the manner in which one parameter is offset against the other so that optimum performance can be achieved needs to be understood whilst VCS is being designed. However, it is not understood by many people as to how the trade-off question needs to be answered from a viewpoint that is scientific.

2.8.2.1. Static Accuracy

The ability of a tracker whereby in space the coordinates of a helmet can determine to be determined is the static accuracy. With the application of no deviation or averaging in a single sample, this figure gives the maximum deviation.

This value is objective as it is a measurement that is the most empirical (raw data value). It is important to both mention the many of measurement of the accuracy as well as the kind of confidence window that has been applied whilst the accuracy is being quoted. Static accuracy is influenced by the certain factors namely: receiver sensitivity, analog to digital converter resolution, rounding error during algorithm computation, algorithm errors, operator errors, transmitter signal to noise ratio, analog component noise tolerance levels, environmental effects, and installation errors.

The error of the overall system is affected differently by each of these factors. The effect of some of the factors on static accuracy is more than that of the others. However, for an overall system error, neither can the individual errors be summed up, nor is it possible to attribute the overall performance of the system to a single component by looking at the maximum error.

2.8.2.2. Dynamic Accuracy

With the movement of the sensor of the tracker, the system accuracy that takes place is the Dynamic accuracy. It is an attribute that is additional to the static accuracy. The factors contributing to the dynamic accuracy of the system are: architecture of the system, type of processor, and components of the system that are time-dependent.

The dependency of dynamic accuracy on static accuracy is quite large. The trade-off between dynamic and static accuracy is quite straightforward when the period used for integration is long. The dynamic accuracy can be quite good over an integration period that is quite long for a large number of acquisitions.

The dynamic accuracy suffers with the application of a time limit for fast-tracking systems, and this is often done. The number of acquisitions shall be fewer when the time available to solve the coordinate position after acquiring data is less.

2.8.2.3. Phase Lag

For a space tracking system, the parameter that is the most important is perhaps the phase lag which is the system's ability to determine the position of the object with a time frame that has been given once the object's position changes.

In a system that is visually coupled generation of scene images in real-time is very important. The visuals that are presented to the user can suffer severe lags that are unacceptable when the tracker system has a phase lag. Simulator sickness is the disease that results due to this. Phase lag is mainly influenced by: the algorithm, type of processor and the architecture.

The choice of all the elements should be for real-time performance and system architecture is very important. At the acquisition stage collection of real-time data is critical, and this becomes all the more applicable at that stage. To ensure that the figure of the overall phase lag is low, the position and orientation of the data should be calculated very fast by the processor.

The kind of algorithm that is used is quite important. The calculation of position and orientation shall have to be done many times by the processor in an iterative solution. The solution that is derived by the iteration shall be more accurate when the algorithm has been designed properly. However, the phase delay increases with every iteration. In another approach for each set of coordinates orientations are calculated through the use of a deterministic solution.

The phase lag performance is affected by all the components that are time dependent. Phase lag reduces with an improvement in any of the components. At the same time, static/dynamic accuracy and phase lag have a trade-off. The requirements of the end application need to be understood when a tracker is to be applied to a VCS, and this is where the trade-off is related.

2.8.3. Helmet-Mounted Display Issues

For a VE system with regard to the display system, a number of parameters need to be considered. The relationship with timing is inherent to these. All the same, display resolution related question shall be raised before these issues are dealt with. The VCS's time criticality issues need not be directly related to this. Sadly though, totally erroneous figures are given due to the abuse of resolution's basic definition by a large number of manufacturers.

When compared to the actual value, the resolution figure that is quoted by them is quite high. This applies mainly where LCDs are the basis of the HMD. Based on a number of elements that are discrete, the LCD manufacturers specify the resolution that is either wrong or right. Red, blue, or green color attributes are given to these elements. A red, blue, and green element comprises of a single-color pixel.

The overall display resolution figure projected by the manufacturers of the devices that are based on LCD is a simple transfer of the resolution figure given by the LCD manufacturer. The individual primary elements comprise a color pixel in the computer graphic and television industry. On a thorough examination, the actual resolution is revealed even though when the manufacturer's display specification is initially examined, it may display a resolution that is adequate.

To take an example, the resolution 208 x 139 pixels (the actual resolution) is quite different from a resolution of 360 x 240 that may be quoted by a manufacturer. The display in a device resulting from low resolution would be quite disappointing when the magnification done is over a FOV that is wide.

Seeing the primary color elements individually is a fairly common thing. Spatial filters or diffuser screens are used over the LCD device by some manufacturers as an attempt to tackle this problem. By this the edge information gets suppressed from the individual primary color elements.

Blurred images resulting in poor displays result from this even though the pixel structure is reduced. Depending upon the task or application, this may/may not be acceptable to the user. The performance of a VCS is influenced by certain HMD parameters namely: Resolution, Exit Pupil, Update Rate, FOV, Eye Relief, and Colhmation.

2.8.4. Image/Graphic Generator Issues

A VCS's performance may be greatly influenced by the design of the graphic system as well as the implementation thereof. In an interactive computer, the problems related to lag are quite complex. Problems generated by lags induced through trackers have been discussed.

Moreover, the manner in which the computer graphic system is constructed can be the base for the overall system lag. High demand is made upon the technologies that are involved in the VE due to the real-time interaction with the VE.

Transmission delays are quoted by very few manufacturers, and the others may not even be able to do the same through the systems they have. The quoting of transmission delays becomes very involved with architectures that are multiprocessor based even though for a single processor system, quantification of the same may be quite a simple process.

In certain cases, a multiprocessor cannot be used to work out the exact delay in transmission, and only with a certain level of tolerance or uncertainty can the timings be calculated. The position lag and the transmission lag are the two kinds of lag that can be defined from the complete system point of view.

2.8.4.1. Transmission Lag

The delay in time that takes place between a user interaction, for instance, the hand or head position tracker, and the movement on the display that corresponds to it is called the transmission lag. The output response shall face a jitter when within the graphics or tracker generator there is any uncertainty or noise in the computation process.

2.8.4.2. Position Lag

Velocity induces position lag and it can be defined in two parts:

- At a given velocity, difference between the distance moved from a position that is known; and
- The distance that is actually moved in the virtual image.

At the same time, position lag is a function of any filtering that is applied to the tracker's output and the velocity of the tracker's sensor. However, position lag and transmission lag are dependent on each other. The effect on a virtual object of the overall position lag and transmission lag can be described simply as a function of real object position which can be given by:

Virtual object position = Real object position – Transmission lag × Velocity.

The calculation and measurement of the parameters is quite involved in practice. For a VCS, the individual lags that make up the overall transmission lag need to be measured so that the system dependent lags can be reduced. The overall system transmission lag is greatly influenced by the architectural design.

Data that is appropriate for a rendering/graphics computer is converted from the tracker data by an input-output (IO) processor that is dedicated for this purpose and to which the tracker is interfaced.

For each eye, identical graphics channels are used. The graphics generators are simultaneously provided with information that the IO processor has generated. Theoretically, the synchronization problems that may exist between the channels for the right and left eye can be removed through this approach.

This architectural arrangement gives certain individual timing delays, and these are given in the figure. The user's head is mounted with a helmet-mounted display in the closed-loop system of a VCS. The feedback path response is affected by the parameters like moment of inertia, center of gravity and mass. In a VCS, the user's overall performance is influenced by this.

Problems like difficulties in synchronization between the channels of the right and left eye can result due to a departure from the idealized VCS. Between the two channels, this is manifested as an apparent difference in the update rate. If more information is displayed on one channel, this defect can crop up even in an idealized VCS, for instance, where a greater number of polygons are displayed.

Prior to the buffers being swopped onto the display generation electronics, the graphic drawing operations need to be written completely into the frame buffers that correspond so that this effect is either eliminated or reduced. For this to occur simultaneously, there must be synchronization of the buffer swop between the two channels.

2.9. COMPARING IMMERSIVE AND NON-IMMERSIVE VIRTUAL REALITY (VR) USER INTERFACES (UI)

Multimedia applications that are sophisticated emerge from asynchronous transfer mode (ATM) technology-based broadband integrated services digital network (B-ISDN) through which bandwidth capabilities are introduced that allows this to happen.

The concept of VPN or virtual private network and logical connectivity is included in the ATM networks. The different sites of a customer are linked together through a set of network resources in a VPN like semi-permanent virtual connections (VPC) and user network interfaces (UNIs). The network management task's complexity increases through this logical connectivity even though more than physical connectivity it provides higher management flexibility.

The same physical links are shared between the operation of different networks; hence there are some dependencies between them with respect to the concept of VPN. Hence, to ensure that the network is effectively managed in real-time, there is a requirement of some kind of collaboration between the carrier and the private networks' network management system.

In comparison to the traditional UI greater interactivity and visualization-ability is required from the emerging networks' management. In these environments, hundreds of ATM switches and virtual channels in tens of thousands have to be dealt with by the manager.

The uses of VR user interface technology for applications in network management have been investigated so that the network management operating environment can be enhanced. 3D user interface technology of two kinds has been introduced in this chapter. First, an immersive VR environment's design and implementation has been explained consisting of certain 3D input devices and HMD.

2.9.1. Immersive VR User Interface

3D navigation and 3D image rendering tools are the basic elements that make up this system. This is coupled with an existing network management system that is based on SNMP (Cabletron SPECTRUM). The network management system gives out three kinds of information namely: performance or fault data, topology, and network configuration. Being dynamic in nature, the fault and performance data require updating on a continuous basis, whereas the others do not require much updating and are nearly static.

With a change of the network topology from the NMS automatically there is the construction of a virtual network world database whereby connectivity information and network configuration are extracted. The VE is built by the VR system using this database.

Fault and performance data needs to be collected from the NMS directly so that a real-time interface can be provided. A direct link is established between the VR systems and NMS so that this can be achieved.

Through this link, the NMS receives inquiries from the VR system so that the required data can be retrieved. The NMS receives the operator's manipulation of network elements as well from the VR system for its application to their real counterparts. The underlying network provides a physical interface between the systems.

The user can fly or walk round the network, navigating through it once the virtual network world has been constructed. The existing network performance levels are represented by the status of the icons. For example, the color represents the operational status, the amount of load that the connection is carrying is represented by the thickness of a link and a broken line represents a disconnected link.

Speech synthesis can be used to present information correlated to the alarm and spatially locating the sounds in 3D can provide clues to the alarm's location. This is different from the WIMP based systems that exist currently in which color-based roll-up procedures and topological maps are used.

It is possible to get more information about the status of the object in the virtual world as they are active, and one can simply walk them so that the internal view can be detailed. The virtual paths within the link can be shown by making the object walk in them where the object is a link.

The interfaces that the element contains can be seen by walking in the object where it is a network element. The status and sub-network of the elements can be seen by walking in when the object is a sub-network.

The network's hierarchical structure can be captured through the walk-in metaphor, and the information presented on the screen is constrained to a level that is comfortable for the network operators.

Data Gloves, joystick, mouse, or a Logitech Cyberman 3D mouse is used for navigation in the world. The manner in which interaction can be done by the operator in a more natural way the interface is being examined. For instance, the simplest way to move or disconnect a network element by grabbing it can be done by using VR gloves and grabbing it with a virtual hand.

Representing the network element in a world that is virtual is another important issue. Special rendering techniques like smooth shading and texture mapping can be used to design a scene in a manner that the operator gets immersed whereby the interface is forgotten by them, and they behave as if it is actually the real world in which they are getting immersed.

2.10. CONCLUSION

The importance of VR systems is increasing every day with the advent of new technologies. There is still a vast potential left in the area of VR that can be used in medical and gaming fields that have exploited this to a very good extent. In this chapter, a major focus has been given to understanding the components of VR that is being used for various applications. In this context, the VR tracking systems has been discussed that provides a better undertaking of how the position of user is tracked while they are using a VR application.

The VR systems can be used for training of new entrants in many fields such as military, medical, etc. This will reduce the cost of training and will have enhanced accuracy. The conclusion that can be made from the chapter is that VR can be used to exploit a feature of a system in a normal scale system. The VR has the ability to perform this action by monitoring, observing, and in a controlled environment. The environments provided by the VR systems is much safer than the real environments, and also there are plenty of chances provided in the VR environment.

REFERENCES

1. T.Ramaprabha, M.Mohamed Sathik, (2012). *The Efficiency Enhancement in Non-Immersive Virtual Reality System by Haptic Devices*. [eBook] *International Journal of Advanced Research in Computer Science and Software Engineering*. Available at: http://ijarcsse.com/Before_August_2017/docs/papers/March2012/volume_2_Issue_3/V2I30082.pdf (accessed on 29 April 2021).
2. Asmaa Saeed Alqahtani, Dr. Lamya Foaud Daghestani, Prof. Lamiaa Fattouh Ibrahim, (2017). Environments and system types of virtual reality technology in STEM: A survey. [eBook] *International Journal of Advanced Computer Science and Applications*. Available at: https://thesai.org/Downloads/Volume8No6/Paper_10-Environments_and_System_Types_of_Virtual_Reality.pdf (accessed on 29 April 2021).
3. Agocg.ac.uk. (n.d.). *Types of VR System*. [Online] Available at: http://www.agocg.ac.uk/reports/virtual/37/chapter2.htm (accessed on 29 April 2021).
4. Bamodu, O., & Ye, X., (2013). Virtual reality and virtual reality system components. [eBook] *Paris: International Conference on Systems Engineering and Modeling*. Available at: https://www.atlantis-press.com/article/5686.pdf (accessed on 29 April 2021).
5. Jones, H., Gigante, M., & Earnshaw, R., (2014). *Virtual Reality Systems*. Kent: Elsevier Science.
6. ResearchGate, (1997). *(PDF) Comparing Immersive and Non-Immersive Virtual Reality User Interfaces for Management of Telecommunication Networks*. [Online] Available at: https://www.researchgate.net/publication/228742087_Comparing_Immersive_and_Non-Immersive_Virtual_Reality_User_Interfaces_for_Management_of_Telecommunication_Networks (accessed on 29 April 2021).

CHAPTER 3

Visual, Acoustic, and Haptic Perception in Virtual Reality

CONTENTS

Visual, acoustic, and haptic are three important senses of humans that impacts the overall experience of VR. In the chapter, visual, acoustic, and haptic perception in virtual reality (VR), the impact of each element in offering the virtual environment (VE) to people is discussed. It explains the visual perception, acoustic perception, and haptic perception of VR.

It also shed some light on the perception deception: audio-visual mismatch in VR using the McGurk effect. The chapter also highlighted the influence of multi-modality on moving target selection in VR. At the end, the chapter explains some of the new innovation that happened in the VR Haptic in which some recent gadgets of VR haptic and its key feature explained.

3.1. VISUAL PERCEPTION OF VIRTUAL REALITY (VR)

It is generally seen that the design of visual presentations for VE is complex because the human visual system is extremely subtle to any anomalies in perceived imagery (Larijani, 1994). The minutest, almost imperceptible anomaly becomes terribly evident when motion is presented into a virtual scene, because visual flow field cues take on an unnatural appearance (Kalawsky, 1993). More precisely, if a VE is unable to produce adequate optical flow patterns, then the user becomes very aware that the experience is unnatural.

One of the major factors is the viewer's visual field when he or she wears a head-mounted display (HMD). There are a variety of ways to graphically present the field of view (FOV) of an individual, embracing equal-area projection plots, polar charts, and rectilinear plots (Kalawsky, 1993).

These manifold tactics make it almost challenging when comparing VE system on the basis of their FOV. Furthermore, manufacturers' estimates of FOV are majority of the time erroneous (Mapes, Rinalducci, Cinq-Mars, and Higgins, 1996; Robinett and Rolland, 1992). Therefore, researchers usually evaluate the FOV of their device independently through optical analyzes or engineering.

Stereopsis is also one of the important elements of the visual modality of VR. Stereopsis defines the perceptual transformation of variations between the two monocular images observed by the eyes. It is a functional constituent of depth perception, however, and not a crucial element for perception of depth.

Stereopsis is imperative to study because of the importance placed on the importance of depth perception to virtual world performance (Ellis and Menges, 1995; Ellis and Bucher, 1994; McDowall, 1994).

The portion of the visual field shared by both eyes is known by the name as the binocular field of vision (i.e., stereopsis) (Haber and Hershenson, 1973). Partial binocular commonality can be used in VR to attain depth perception, in which a monocular image is relocated outwards or inwards. Such partial overlap can be taken into usage to recognize wide FOVs with smaller and lighter HMDs.

In a way to have effectual depth perception in VR with partial overlap, although, the needed degree of overlap must be computed. Presently, the sum of overlap needed is not comprehended fully. Human-factors practitioners are required to engage in perceptual and human performance studies to find out if partial overlap is satisfactory and what levels are needed for distinctive applications (see Mon-Williams et al., 1993; Rushton et al., 1994).

3.2. ACOUSTIC PERCEPTION OF VIRTUAL REALITY (VR)

In order to make a realistic auditory environment, it is imperative to attain a superior comprehending of how the ears receive sound, chiefly concentrating on 3-D audio localization. Audio localization supports listeners in differentiating distinct sound sources.

Localization is mainly identified by intensity alterations and phase or temporal differences between signals at the ears (Middlebrooks and Green, 1991; Begault and Wenzel, 1993). More explicitly, the range up to which one sound masks another relied on the comparative intensity, frequency, and location of sound sources. Audio localization is impacted by the existence of other sounds and the track from which these sounds is being produced.

The human ear can detect a sound source even in the existence of strong opposing echoes by refusing the unwanted sounds.

> According to Biocca (1992a), it was stated that "the aural realism of virtual spaces requires replicating the spatial characteristics of sounds like the changing intensity of a race car engine as. it screeches past."

The human ear also has the potential to segregate a specific sound source from among a group of all others producing from distinctive locations (Koenig, 1950). In order to efficiently advance aural displays, this listener ability to focus and track in on a specific auditory source (i.e., the cocktail party effect) are required to be understood in a better and efficient manner.

Auditory localization can be comprehended in the horizontal plane (left to right). Sounds can arrive 700 microseconds earlier to one ear in comparison to other, and the sound in the further ear can be lessened by as much as 35 decibels compared to the nearer ear (Middlebrooks et al., 1991).

For instance, if a listener senses a sound coming from the right corner, usually the sound has been received by the right ear firstly and/or is therefore louder in the right ear in comparison to the left. When sound sources are afar one meter from the head, these intensity differences and interaural time become less pronounced in supporting audio localization.

It is generally seen that vertical localization in the median plane cannot be contingent on interaural disparities (for instance, as long as there is symmetry between head and ear). When a sound is unswervingly in front of (or behind) a listener, the interaural differentiations are 0; although, the listener is still able to somehow localize the sound.

In these scenarios, the structure of the external ear (i.e., the pinna) is believed to harvest variations in the spectrum of a sound (i.e., spectral shape cues) that support in localizing the sound (Middlebrooks et al., 1991; Fisher and Freedman, 1968).

Thus, in a way to efficiently characterize 3-D audio localization, binaural localization cues which are obtained by the ears can be signified by the pinna cues, in addition to by the interaural acoustical differences.

Currently, a head-related transfer function (HRTF) has been taken into usage to characterize the manner in which there is changes in the sound sources with the change in the position of head by a listener and can be definite with awareness of the source position and the orientation and position of the head (Cohen, 1992; Butler, 1987).

The HRTF is primarily relied upon on the physiological makeup of the ear of the listener (for instance, the pinna does a nonlinear fitting job in the HRTF). Current advancement has provided the way for the development of personalized HRTFs (Crystal Rivers Engineering, 1995). These tailored tasks still need a huge amount of calibration time.

Preferably, a more generalized HRTF could be formed that would be pertinent to a multitude of users. It can be viable since the transfer functions of the external ear have been identified to be akin across distinctive individuals, although there believes to be a downward shift in spectra frequency with growing physical size (Makous, and Green, 1989).

3.3. HAPTIC PERCEPTION OF VIRTUAL REALITY (VR)

A haptic sensation (i.e., touch) is a mechanical contact with the skin. It is imperative to integrate haptic feedback in VE because such feedback is believed to greatly improve performance (Figure 3.1) (Burdea et al., 1994).

Figure 3.1: Haptic perception of virtual reality.

Source: Image by Teslarati.com.

For instance, by merging haptic feedback into synthetic molecular modeling systems, the problem-solving potential of chemists was improved considerably (Batter and Kilpatrick, 1990; Brooks et al., 1990).

There are basically three mechanical stimuli that create the sensation of touch: a dislocation of the skin over an extended period of time, a transitory shift of the skin that is recurrent at a variable or constant frequency (Geldard, 1972), and a transitory (a few milliseconds) displacement of the skin.

Even with the comprehending of these global mechanisms, although, the attributes of the skin are almost challenging to depict in a quantitative fashion. It is primarily because of the fact that the skin has variable thresholds for touch (vibrotactile thresholds) and can execute temporal summations and complex spatial that are all a purpose of the position and type of the mechanical stimuli (Hill, 1967). So as the stimulus varies, so does the sensation of touch, thus initiating a challenge for those aiming to model synthetic haptic feedback.

Another haptic issue is that the sensations of the skin acclimate with disclosure to stimuli. More exactly, the sensation effect lessens in sensitivity to a continued stimulus, may disappear entirely even though the stimulus is already existing, and differs by the type of receptor.

Phasic receptors are ones that adapt rapidly and associate with touch, pressure, and smell. Tonic receptors that are correlated to body position and pain gradually adapt and may have an afterimage that stays even once the stimulus is eradicated.

Surface characteristics of the stimulus also impact the overall sensation of touch. For a hard surface to be felt after initial contact, there should be maintenance of active pressure. The sensation of textured surfaces needs some comparative motion between the skin and the surface to be maintained.

Soft surfaces can employ and ensures a slight positive reaction against the skin after the primary contact without relative motion or active pressure. Most present systems offer restricted haptic feedback, mainly isolated to the hand.

For instance, vibrotactile and electrotactile devices have been created to pretend the sensation of texture as well as illusions of other surface on the hand (Webster et al., 1991). There may be a requirement of more fully haptic feedback in order to improve the presence and performance (Biocca, 1992a). Current exoskeletons are expensive and clumsy. A less-invasive means of offering full haptic feedback is preferable.

In order to convey the sensation of synthetic remote touch, it is thus imperative to have full knowledge of the mechanical stimuli which harvest the sensation of touch, the effect of a sensation, the vibrotactile thresholds, the adaptation of these receptors to some sorts of stimuli, and the dynamic variety of the touch receptors.

Thus, the human haptic system is required to be more wholly described, possibly through a computational model of the physical characteristics of the skin, in a way to produce a synthesized haptic response.

3.4. PERCEPTION DECEPTION: AUDIO-VISUAL MISMATCH IN VIRTUAL REALITY (VR) USING THE MCGURK EFFECT

It is generally seen that VR system puts a participant in a computer-generated 3D environment that expands or mimics the real world. In the VR research, the main area of research is to think how to make it better by providing an immersive experience. Humans immersive experience are basically the result of multisensory integration (Figure 3.2).

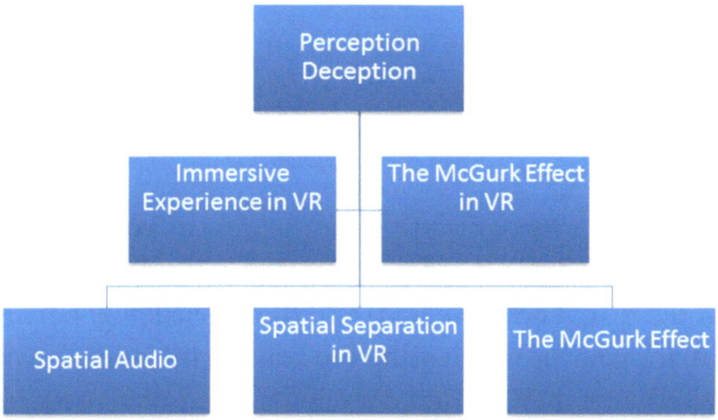

Figure 3.2: Perception deception.

VR researchers are in search of developing technologies that can offer precise sensory cues delivering increasingly natural sensorimotor contingencies, raising the probability of virtual events. The sensory cues in state-of-the-art VR systems (e.g., Vive, Oculus) chiefly emphasize on auditory and visual immersion with some haptic feedback in hand controllers.

Although the technology being used in the creation of immersive visual content has become more powerful, affordable, and accessible, focus on creating auditory content and synchronizing the audio-visual has lagged, impacting the overall fused immersive experience. In order to provide the immersive feeling to the user, there should be a proper match between the vision and audio.

If there will be no coordination between the sounds and audio cues aligned with the associated visual experience, the resultant incongruity instigates the virtual immersion to collapse. Multi signal mismatch is one of the most prevalent problem because of the concerns related to latency, headphone, head-tracking technology, and immersive media content production.

Despite the possibility of breaking the immersive feeling because of inevitable multisensory disparity, an exciting phenomenon saying that humans can recognize a united precept in the occasion of incongruent stimuli first presented by psychologists McDonald and McGurk in 1976 stimulates new concerns of audio-visual design for VR.

This phenomenon suggests that human minds can join together off-balance sensory information to form a satisfactory conclusion about an experience, and being learned using a classical research tool known as the McGurk effect experiment.

The McGurk effect recognize the significance of knowing the abilities of the unisensory stimuli (i.e., the determination of auditory components and the lucidity of the visual components) and the consistency between the input sensory information before evaluating the general integrated experience. The immersive experience in VR often depends upon multisensory integration.

In this section, the major area of focus would be on spatial qualities of audio-visual stimuli, particularly the association between where sounds are seeming to be coming from comparative to their analogous visual source. A comprehensive thoughtful of the significance of image resolution and spatial audio localization on the immersive experience can direct the growth and function of compression from a quality of experience perspective.

3.4.1. Immersive Experience in VR

It is generally seen that Immersive experience is an outcome of multisensory processing. Sensory inputs for immersive experience usually comprise of audio, vision, force feedback and touch and less often taste and smell. In the setting of VR, the ultimate objective is to trigger as many human natural sensory inputs as probable via computer-based technology (Figure 3.3).

Figure 3.3: Immersive experience in VR.

Source: Image by science news for students.

The imperative tools advanced to attain this objective comprises of stereo, wide field of view vision, low-latency from head move to display, head tracking, and high-resolution displays. Although, because of the multisensory property of the immersive experience, a single advancement in one phase is not adequate to enhance the overall immersive experience in VR.

When a virtually produced multisensory illusion gives an impression of a high plausibility, it is hence assumed that the designed scene is really occurring. Any disparity between different senses would make a scene less impressive and ensuing in the immersive sense of existence in a virtual scene collapse.

3.4.2. Spatial Audio

In practice, designers have been widely making use of 3D sound techniques to improve the scenes covered in the auditory information to address the challenges of the traditionally-used stereo recordings. Although, quality 3D audio is technically complex to transport for a number of reasons. These include compromises in present content capture, formation, and distribution.

For instance, many prevailing economical 360◦ cameras still make use of stereo or mono as their audio signals. Without extra effort incurred in rendering and capturing the spatial audio, the camera-recorded audio-visual content is incompetent of generating a truly immersive sense of being there in the VR environment.

Enthusiastically updating both the auditory and the visual signal on the basis of orientation poses and head-position further challenges to both compression and network technologies used to bring immersive VR.

3.4.3. Spatial Separation in VR

The discrepancy between modalities can happen as an outcome of a timing disparity between tracking and network technology. It may also be because of a mismatch of localization information as a result of rendering technology and audio production. A comprehensive conversation of network and hardware factors that can result in differences is beyond the possibility of this chapter. This chapter will emphasize on the insight of spatial separation.

The consequences of audio-visual spatial separation are still not certain. A misalignment of visual and auditory signals can adversely affect VR immersive experience. Although, psychologists recognize the importance of the human mind to form illusions given dissimilar stimuli.

For instance, the 'ventriloquism effect 'says given a visual effect, the location of the auditory sources can be discounted, so creating an illusion as if the sound source is coming from the identical place as the visual.

A related idea was also imitated in the 'unity assumption' effect. These theories trigger the incentive to scrutinize whether spatial separation is perceptually negative in VR. The degree up to which spatial separation can be tolerated will be significant to both VR content and technology developers.

3.4.4. The McGurk Effect

To examine whether a fusion effect occurs in the occasion of visual-audio spatial separation in a VR context, this study is based upon a classical research tool known as the McGurk effect that discovers a phenomenon of a changed perception of auditory speech signal given dissimilar audio-visual pairing. It is selected because of the potential of the McGurk effect to 'reveal the potential of visual-audio integration.' In a traditional McGurk effect experiment, a participant is generally offered with a visual and an auditory signal concurrently where each signal conveys distinctive speech tokens (i.e., audio of /ba/ together with a vision of /ga/).

The participant's stated subjective auditory precept is anticipated to diverge from what is truly offered acoustically as an outcome of unaware integration of phonetic information (i.e., hearing /da/ when being presented with visual /ga/ and audio /ba/).

Such categorical change of speech perception is portrayed as the McGurk effect. Since the early publication of this lab-controlled illusion experiment in 1976, the McGurk effect has appealed to a mass in the arena of cognitive science.

Psychologists are of the view that the 'spatial proximity' and 'laws of common fate' are the two fundamental theories for this phenomenon. These theories are dependent on the fundamental principles of perceptual information fusion. Further down the line, researchers have been aggressively examining factors that provide the contribution to this phenomenon.

Aside from the age factors and interpersonal difference, those factors unfolding stimuli comprising talker voice, visual degradation, time lag in synchronization, and choice of utterance, all have been confirmed to have an effect on the potential of McGurk effect. All these inspiring factors lead to a further generalization of the conditions for the McGurk effect to arise: the consistency of the manifold sensory signals and the quality of the unisensory signal.

3.4.5. The McGurk Effect in VR

The spatial separation of the visual and the audio information is one of the most significant challenges in the VR content creation process. Based on the existing findings of the conditions for the incidence of the McGurk effect, factors comprising the quality of the visual information, the quality of sound localization information, and how far apart the separation between the auditory and visual scenes are all vital concerns to offer or facilitate users with an immersive experience.

Typically, the sound localization information should be designed in the extreme comprehensive and correct manner to balance the visual scene so that the combined effect can produce a higher plausibility (Psi) to pretend a participant into thinking that the virtual scene is actually happening. Meanwhile, the 'ventriloquism effect, which is akin to children where a puppet appears to speak, implies some tolerance of the coherence of the multisensory signals for the fusion effect to take place.

However, the existing literature offers inconsistent conclusions with respect to the consequences on the combined speech perception in the event of visual-audio disparity. There was no clear effect located on the audio-visual integration when the spatial separation was up to 37.5°. Although, a weak separation effect was discovered when the separation angle reached to 60°.

According to Jones and Munhall, there was the meager difference in the effect on the McGurk effect when mounting the spatial separation of the visual and the auditory scenes. Siddig et al. Also attained these results with the help of 3D audio virtually binaurally rendered spatial audio for up to 90°. According to Jones and Jarick, it was concluded that the spatial separation impact was only imperative when the sound was 180° away from the visual.

All of the experiments conducted above was based on 2D visuals only, although there was use of spatial sound in order to produce the auditory signals. In this, the main area of interest will be on inspecting whether the existence of the fusion phenomenon would alter in a 3D scenario where participants are positioned in a surrounding environment aroused by both 3D visuals and 3D audio. As in a 3D environment, a participant has a hope of the source of the audio with respect to the visual stimuli, and in this case, the mind is highly projected to fuse multisensory signals to pretend them into believing that a scene is real. Therefore, it is suggested that an alteration of visual scenario from 2D to 3D could lead to an alteration in the integration effect. Following the existing reasoning of the McGurk effect discussed in earlier section, it is sensible to hypothesize that the potency of the McGurk effect in a VR context are an outcome of resolution of the visual signals, articulacy of speech, and varied range of spatial differences between the sources of the visual and auditory signals.

If the psychological theories with respect to the perceptual phenomena believe to be right in a context of a VR, one can endure an amount of audio-spatial separation resulting from digital and streaming content processing as it will not cause a negative integrated experience. Building on the experiment defined earlier, the ultimate objective is to offer an insight on multisensory data compression factors to deliberate for VR streaming applications.

3.5. INFLUENCE OF MULTI-MODALITY ON MOVING TARGET SELECTION IN VIRTUAL REALITY (VR)

It is important to note that the moving target selection is a fundamental task that picks the moving targets in a 3D space or in a planner. In a kind of moving target selection task, the users are required to place the pointer in the target space, without choosing the target, in a static time window (Figure 3.4).

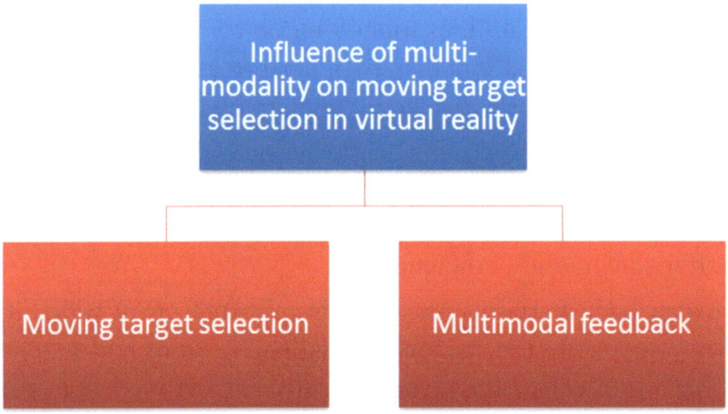

Figure 3.4: Influence of multi-modality on moving target selection in virtual reality.

The performance of user is significantly impacted by the size of the moving target and the moving speed. In a second kind, the users require to finish the target selection in a static time window that is not confined by the spatial location of the target. It is usually seen that selection of target either too late or too early could lead to failure. Another moving target selection task needs spatial accuracy apart from the accuracy in timing, consistency, and rhythm.

Actual dynamic interaction tasks are generally more intricate and are usually impacted by several factors such as space, time, environment, and modality cues. Because of the universality of the task, moving target selection has been used widely in music, games, teaching, as well as other fields. The moving target selection task has been studied in particular time and space areas.

There are two chief sorts used to enhance the precision of selection in moving target selection: cursor enhancement and target enhancement. The primary objective is to enhance interaction accuracy by raising the optionality of time and space.

According to the cue perception fusion theory, multimodal cues have a huge impact on user motion perception. It is widely supposed that the superposition of multimodal cues (such as auditory, vision, and haptic) that carry steady information can improve user perception accurateness and improve the performance.

Although, in a VR environment, distinctive modality cues have distinctive expressive capabilities to moving objects. One key question that is prevalent in this case is whether user performance in multi-perception modality is better than that in the single perception modality.

The comparison between the trimodal and the bimodal conditions also requires to be studied. It is of high importance to study the effect of complementary or unnecessary information of a distinctive perceptual modality on the interaction efficiency.

VR scenes have certain prevailing issues such as vertigo, diplopia, and lack of perceptual cues, which are distinctive from real scenes. Users generally have ineffective interaction because of the inadequacy of some cues.

Conventional VR systems typically engage a combination of auditory and visual modality to enhance the target selection accuracy. With the popularity of handheld mobile devices, data gloves, and other tools, users have begun learning haptic information to improve user experience, enrich interactive information, and advance interaction efficiency.

Earlier, researchers have a major area of focus on the effects of distinctive modalities on static 3D target selection. In this study, we take into consideration the impacts of distinctive interaction modalities on user performance and subjective rating of 3D moving target selection in a VR environment.

The chief contributions are as follows: first, attaining a set of empirical data of multimodal conditions affecting the stirring target selection in VR; second, examining the results of target speed, modality combination, and target angle on subjective rating and user performance of the moving target selection; third, obtaining, and summarizing the outcomes of a set of multimodal conditions on the stirring target selection in a VR environment. Strategies for the interface design are offered for future interaction design in this field.

3.5.1. Moving Target Selection

While keeping an eye on the trajectory of a moving target, moving target selection requires to propose the time of target selection that needed high coordination potential of user motion-perception. Thus, the motion ability and the level of perception will impact the error rate of tasks and overall achievement time.

Jagacinski et al. exhibited that the target moving speed has a substantial impact on the accomplishment time of the target selection task. The quicker the target moving speed, the higher challenging it is to select a user, and the time consumed is lengthier.

Zhai et al. deem that the primary size of the target does not impact the overall target selection time, while the target size will impact the overall time required for the target selection when the user clicks on it.

Huang et al. have indicated that the endpoint distribution for a 1-dimensional moving target selection is significantly impacted by the moving speed as well as the width of the target, although, not by the preliminary distance between the target and pointing device Lee et al. reviewed the selection of the moving targets in a static time window.

There are basically two factors that are related with the error rate of the selection: the time window width in which the targets can be chosen and the overall waiting time length for the moving targets. To elucidate the effect of visual factors on the election results, and augment the practical use of forecast results, Lee et al. further deliberated the effect of temporal structure (rhythm, repeatability, etc.), and visual perceptible clues (color, moving target speed, etc.), on the forecast accurateness based on the previous work.

To lessen the impact of size and speed on the error rate and the completion time of the moving target selection in distinctive scenarios, a multiplicity of methods to surge its interaction efficiency are planned, including lessening the distance from the target and enhancing the size of the target.

For instance, Area Cursor proposed by Kabbash et al. wishes to enhance the contact range between the target and the cursor and lessen the moving distance between them by enhancing the cursor area, Bubble Cursor which is invented by Grossman et al. makes the cursor comprise only one selected target in the selected range through the adjacent characteristics of the target, and recovers the overall completion time of the user selection by gradually changing the cursor selection range.

Hasan et al. added the selected area which is akin to the comet tail to the target according to the target speed and the moving trajectory. The tail size is allied to the target moving speed to enhance the selected area of the target.

3.5.2. Multimodal Feedback

Multimodal human-computer interaction (HCI) is basically the interdisciplinary field of computer vision, psychology as well as several other areas. It is extensively used in the field of HCI. In real scenes, users mainly interact with the outside environment only with respect to the information attained by visual cues, which enhances the load on the visual system and lessens the information attainment potential of human auditory and haptic sensory systems.

It is generally seen that users are impacted by distinctive perception modalities under distinctive information resources and distinctive interaction conditions, are obtained by using multi-perception modalities.

For instance, when the user judges the shape, the size, and the location, the results are more vulnerable to visual cues when offering the user with visual-haptic modality cues. Although, for some application situations with weak visual modality cues, their judgment is highly vulnerable to the impact of haptic modality characteristics.

Choosing an adequate multimodal feedback can enhance the robustness of the system and lessen the error rate of interaction. Andy and Stephen learned the impacts of multi-modality on the small target selection in GUI, including tactile, non-speech audio, and viscous pseudo-haptic feedbacks, and their combinations on fixed target attainment tasks, to direct designers in selecting distinctive modalities to surge the overall experience of target selection.

3D target selection in VR as a basic task is one of the most widely discussed topic with the development of the 3D user interface. The data acquisition preciseness by tracking devices and the perception information offered by the modality feedback restricts the efficiency of actual interaction.

Thus, several methods are projected to progress the subjective feelings and user performance of target selection. To lessen the restriction of tracking device correctness, scholars have learnt the relationship cognitive function and between gesture pointing motion Schmidt offers a pointing-based probabilistic selection algorithm to lessen tracking insecurity.

Otherwise, the shortage of modality information or the distinction between the real-world cues and the modality information will lessen user operation performance. Visual cues are generally taken into usage to instigate users while picking targets in a VR environment, for instance, if the target is picked, there would be a change in target color. Though this method lessens the redundancy of information, it does not recuperate the performance of user interaction, and may even lessen the efficiency of user selection.

Haptic and auditory modalities are generally used as auxiliary perception modalities to enhance the user's subjective experience and interaction efficiency. Menelas et al. aided users to select and locate the targets closed in VEs with auditory or haptic cues. Cabreir et al. learnt the impacts of multi-modal cues on the interaction efficiency of static target selection using mid-air gestures by elderly users.

Ariza et al. examined the effects of continuous and discrete cues of multimodal feedback on static target selection. Faeth et al. added visual, auditory, and haptic feedback to virtual buttons, and reviewed the effects of distinctive modalities and their groupings on the error and the time rates of virtual button tasks.

In the VR environment or the real world, although there are several research results on the effect of multi-modality on static target selection, few have been deliberated the impact of modalities on the result of moving target selection. Mould et al. reviewed the visual cues influence on the finished time and the error rate of target selection under distinctive occlusion conditions in a 2D graphical user interface.

The target was blocked by other objects whose moving speed and number varied in several ways, comprising no visual feedback, visual feedback only for certain objects, and visual feedback for all objects. In this study, the major area of focus will be only on 3D user interface in VR.

In this, we review the impact of visual modal cues on moving target selection, apart from the influence of other modalities on subjective feelings and user performance, which can be taken into usage to guide the interactive design of the moving target selection in a VE.

3.6. WHAT'S NEW IN VR HAPTICS?

It is well known that VR is evolving at an astounding rate. Some of the humanity's most exhilarating technologies and tools are coming to the VR space. One such technology that is making VR more powerful is the VR haptics technology (Figure 3.5).

Figure 3.5: VR haptic technology.

Source: Image by IEEE Spectrum.

VR Haptics technology provides an additional dimension to the VR world by allowing users to actually feel the VE via the sense of touch, in addition to aural and visual perception. It makes a person feel truly immersive in the artificial world. For instance, it has the potential to make a person feel that he is in a desert seeing the sand, in addition to feeling it glide under their feet as they walk around.

It uses external devices such as Shoes, Gloves, Joysticks, etc., through which users can collect feedback in the arrangement of vibrations from these computer applications. This feedback offers physical sensations in hand as well as to other parts of the body. It also offers a realistic simulation of the behaviors and movements comparable to those grasped in the actual world.

3.6.1. VR Haptics: A Growing Domain

The VR haptics technology is developing beyond generating vibrations in game controllers. Now, in the coming few years, one can expect to cuddle a dog and feel it licking your face in the VR world. This gives an indication about the speed at which the haptic technology is developing.

One good example that considers modern VR is the well-liked sci-fi novel "Ready Player One." It demonstrates the potentials of haptic technology in the future. The novel investigates the trip of a guy as he sets foot into a VR simulator (OASIS). He wears a pair of gloves and a headset to move around the virtual world.

In addition to the gloves, a variety of future concept products are also covered in the novel, which makes the fantasy of immersion simpler to portrait, such as Temperature/ Wind generators that mimic real-life and towers producing smells in the VR world.

The introduction of haptics was almost similar to head-mounted displays (HMD), which came into the limelight in the year 2010s. HMDs enable the opportunity for people to see the VR while haptic feedback offers the potential to experience the virtual world and to act within it.

Temperature, texture, taste, pressure, smell, and other non-visual sensory inputs can be perceived by the individuals in the VR. Apart from VR apps and games, Haptics feedback is widely used in personal computers (PCs), robots, mobile devices, and more.

But, in this section, the main area of focus will be only on the use of haptic feedback or haptic technology in the VR space. Generally, majority of the VR users make use Touch Controllers for haptic feedback. But, in the recent times, a large number of third-party companies are jumping into this market by launching their product such as gloves for systems like the HTC Vive and Oculus Rift.

Here is a shortlist of current progresses in the haptic technology for the VR world:

- ***Super Affordable VR Haptic Gloves by Plexus:*** Most of the presently viable options in the VR haptics field are somewhat pricey, but some of them are worth trying, such as a recently introduced by the company Plexus, a VR haptic and sensor glove.
  - Key Features:
    - Plexus VR haptics gloves provide an entirely modular tracking solution which has the potential of tracking up to 0.01° of accuracy.
    - These gloves have the potential of single finger tracking as well as tracking individual joint on the

finger, thus, providing higher accuracy in the VR world.

- It is compatible with the Oculus Rift, HTC Vive, as well as Windows mixed reality (MR) devices.
- It is also important to note that the VR haptic gloves come with extra adapter plates.

The development kit version of the Plexus haptic gloves, which is priced at $249 per pair of gloves, can be pre-ordered on the official Plexus Website. It is available in several companies such as USA, Canada, Europe, and Australia.

- ***Kaaya Tech's Full Body Tracking HoloSuit:*** Another example of latest haptic technology is the kaaya tech's full-body HoloSuit. Kaaya came out with a motion capture (MoCap) suit known as HoloSuit, which is enabled with the technologies of MoCap as well as haptic feedback.

It is worth noticing that HoloSuit is the world's first most economical wireless, bi-directional, easy to use, full body MoCap suit. HoloSuit has the potential to capture user's whole body movement data and uses haptic feedback to transmits information back to the user.

- Key Features:
  - The HoloSuit comprises of 36 embedded sensors in the pro version and 26 embedded sensors in the less intricate version. Embedded sensors convey out all the tasks of capturing body motion which is imperative for world-scale tracking.
  - It also comprises of 9 haptic feedback devices, and 6 embedded firing buttons (buttons that is responsible for monitoring specific tasks such as pausing, or saving the game, etc.), which are disseminated across both legs, arms, and all the 10 fingers.
  - It offers data wirelessly either through Bluetooth LE or Wi-Fi to a VR setup by using a Wi-Fi SDK or Unity.
  - It is important to note that the HoloSuit does not come with any external tracking option via camera.
  - It is supported by the majority of large platforms such as macOS, Windows, Android devices and iOS.

A whole HoloSuit is quite expensive and starts at a fixed price of $999. While only Jersey and Jacket are priced at $499, track pants or jersey for $399, and a pair of gloves are offered for $799. However, the HoloSuit Pro is the most expensive one that is priced at $1,599.

- ***Disney's VR Haptic "Force Jacket":*** Another example of the Haptic technology instrument is Disney's VR Haptic "Force Jacket." Disney came out with their VR haptic jacket, widely known by the name, "Force Jacket" few years back. It enables users with accurately directed force in addition to the high-frequency vibration that can be experienced against the user's upper body part in sync with the visual medium. The prototype is created out of a transformed life jacket and is offered with 26 airbags.
  - Key Features:
    - The Haptic Jacket uses a vacuum pump and an air compressor. These air compartments in the jacket can be magnified to exert a force on the body of user comparative to force-sensitive resistors.
    - 26 air compartments are started using microcontrollers for either vibrotactile feedback or pressure or both.
    - Controllers are often taken into usage to activate the solenoid valves that are linked to the vacuum.
    - There are several Jacket inflation parameters such as force, speed, and duration which are quantified using the haptic effects editor.
    - The jacket generally makes use of the motion interface in order to successively inflate the compartments for pretending motion across the body.

Each airbag within the haptic jacket can be induced to mimic sensations such as getting tapped on the shoulder, being hit by a snowball in the chest, getting punched in the side, lime dripping onto the back of a user, and a snake coiling its body around the user.

The jacket is often taken into usage by the gaming and entertainment industry and is not presently not accessible for the consumer market. But it believes to have the huge potential in the future for other applications as well.

- ***VR Gloves by HaptX:*** HaptX's pair of VR gloves is one of the most important innovation in the segment of haptic VR. This pair of VR gloves use micro-pneumatics technology for comprehensive force and haptics feedback (the potential to constrain movement of fingers to pretend holding objects) in the fingers.
  - Key Features:
    - It comprises of technology that enables it to offer 100 points of tactile displacement feedback.
    - It proposes up to five pounds of endurance per finger.
    - It also comes with the technology of sub-millimeter precision motion tracking.
    - The glove uses SDK of HaptX's design, which is generated by using Unreal Engine's physics system. This states the glove where and when it requires to apply haptic effects as well as how and when to employ the force feedback.

However, the company has not yet released any information about the pricing or worldwide availability released to date. But it is supposed to introduce the VR gloves for the consumer market in the next few months.

Apart from these products, there are also other trivial progresses that keep occurring in the VR haptics space. For instance, Heather Culbertson, Assistant Professor of USC's computer department, just made a haptic armband that has the potential of mimicking the sensation of a human touch.

VR aimed at providing an environment where one feels truly immersive and where one can experience the objects as in the real world. These products are getting the VR world a step closer to attain higher levels of immersive experiences.

Those days have passed when haptic feedback was confined to just vibrating joysticks and controllers. With the advancement in the technology, the entire new world of VR haptic devices is here to make the user's VR experience as seamlessly immersive as conceivable. In fact, some people have opined that without Haptics, VR is nothing but a combination of a sound and picture.

3.7. CONCLUSION

At the end, it is concluded that VR is dependent on three human elements such as visual, acoustic, and haptic. Understanding of these is very crucial in order to create a better VE for people. These three factors impact the overall experience of people in VR. If their disparity in anyone, then it can disrupt the whole experience.

The technology designed for VR should take into consideration all these factors in order to efficiently integrate all these elements in VR. The VR technology is continuously improving, and in the future, one can expect significant advancement in this technology. It will likely to address some of the challenges that are still prevalent in the VR and that act as hindrance in smooth experience of VR.

REFERENCES

1. T. Koritnik, A. Koenig, T. Bajd, R. Riener, M. Munih, (2010). *Comparison of Visual and Haptic Feedback During Training of Lower Extremities*. [eBook] ScienceDirect. Available at: https://robolab.si/wp-content/uploads/Publication/Pages/2010/10_Koritnik_GaitPosture.pdf (accessed on 29 April 2021).
2. D. Rausch, d L. Aspöck, T. Knott, S. Pelzer, M. Vorländer, T. Kuhlen, (2012). *Comparing Auditory and Haptic Feedback for a Virtual Drilling Task*. [eBook] Available at: https://vr.rwth-aachen.de/media/papers/ComparingAuditoryAndHapticFeedback.pdf (accessed on 29 April 2021).
3. Altinsoy, M., (2021). *The Quality of Auditory-Tactile Virtual Environments*. [eBook] Dresden University of Technology, Chair of Communication Acoustics, Dresden, Germany. Available at: https://tu-dresden.de/ing/elektrotechnik/ias/aha/ressourcen/dateien/professur/publikationen/Altinsoy2013a_-_The_Quality_of_Auditory-Tactile_Virtual_Environments.pdf?lang=en (accessed on 29 April 2021).
4. Alur, A., Shrivastav, P., & Jumde, A., (2021). *Haptic Technology: A Comprehensive Review of its Applications and Future Prospects*. [eBook] Available at: http://ijcsit.com/docs/Volume%205/vol5issue05/ijcsit2014050508.pdf (accessed on 29 April 2021).
5. Bermejo, C., & Hui, P., (2021). *A Survey on Haptic Technologies for Mobile Augmented Reality*. [eBook] Hong Kong University of Science and Technology. Available at: https://arxiv.org/pdf/1709.00698.pdf (accessed on 29 April 2021).
6. Gutiérrez, D., & Masia, B., (2021). *Multimodality in VR: A Survey*. [Online] ResearchGate. Available at: https://www.researchgate.net/publication/348648114_Multimodality_in_VR_A_survey (accessed on 29 April 2021).
7. Hulusic, V., Harvey, C., Debattista, K., Tsingos, N., Howard, D., & Chalmers, A., (2020). *Acoustic Rendering and Auditory-Visual Cross-Modal Perception and Interaction*. [eBook] Available at: http://eprints.bournemouth.ac.uk/27388/1/PostPrintRenderingandCrossMod.pdf (accessed on 29 April 2021).
8. Kuschel, M., (2018). *Visual-Haptic Presence Systems: Utility Optimization, Compliance Perception, and Data Compression*. [eBook] Available at: https://mediatum.ub.tum.de/doc/654293/681528.pdf (accessed on 29 April 2021).

9. Li, Y., Wu, D., Huang, J., Tian, F., Wang, H., & Dai, G., (2019). Influence of multi-modality on moving target selection in virtual reality. *Virtual Reality and Intelligent Hardware, 1*(3), 303–315.
10. Madhawa, S. P., Pappas, T. N., Atkins, J., West, J., & Hartmann, W., (2021). *Acoustic-Tactile Rendering of Visual Information*. [eBook] Available at: http://users.eecs.northwestern.edu/~aps220/publications/HVEI2012_Silva.pdf (accessed on 29 April 2021).
11. Mathur, N., (2021). *What's New in VR Haptics? | Packt Hub*. [Online] Packt Hub. Available at: https://hub.packtpub.com/whats-new-in-vr-haptics/ (accessed on 29 April 2021).
12. Mihelj, M., Novak, D., & Begus, S., (2013). Haptic modality in virtual reality. *Virtual Reality Technology and Applications*, 161–194.
13. Sagardia, M., & Hulin, T., (2017). *Multimodal Evaluation of the Differences Between Real and Virtual Assemblies*. [Online] Elib.dlr.de. Available at: https://elib.dlr.de/113931/2/Manuscript_TH-2017-01-0007.pdf (accessed on 29 April 2021).
14. Siddig, A., Wenyi, S. P., Parker, M., & Hines, A., (n.d.). *Perception Deception: Audio-Visual Mismatch in Virtual Reality Using the McGurk Effect*. [eBook] Available at: http://ceur-ws.org/Vol-2563/aics_18.pdf (accessed on 29 April 2021).
15. Sigrist, R., Rauter, G., Wolf, P., & Riener, R., (2012). *Augmented Visual, Auditory, Haptic, and Multimodal Feedback in Motor Learning: A Review*. [Online] link.springer.com. Available at: https://link.springer.com/article/10.3758/s13423-012-0333-8 (accessed on 29 April 2021).
16. Sinclair, M., Ofek, E., Gonzalez-Franco, M., & Holz, C., (2019). Capstan crunch. *Proceedings of the 32nd Annual ACM Symposium on User Interface Software and Technology*.
17. Stanney, K., Mourant, R., & Kennedy, R., (2021). *Human Factors Issues in Virtual Environments: A Review of the Literature*. [eBook] Kay M. Stanney, University of Central Florida, Department of Industrial Engineering and Management Systems, Orlando. Available at: http://web.mit.edu/16.459/www/Stanney.pdf (accessed on 29 April 2021).
18. Stevens, J., Kincaid, P., & Sottilare, R., (2015). Visual modality research in virtual and mixed reality simulation. *The Journal of Défense Modeling and Simulation: Applications, Methodology, Technology, 12*(4), 519–537.

19. The National Academic Press, (2019). *Haptic Interfaces*. [Online] Available at: https://www.nap.edu/read/4761/chapter/8 (accessed on 29 April 2021).

CHAPTER 4

Virtual Reality Vs. Augmented Reality

CONTENTS

The following chapter illustrates about the scale of differences between the virtual reality (VR) and the augmented reality (AR). It discusses about all the components of the AR and along with this all the applications of AR in every industry. It illustrates various pros and cons of VR and AR how it is benefit for each single industry from education to entertainment. Further, it highlights the where the AR and the VR is efficient use and how it is important for future. It also tells that with the invention of AR things become much convenient in every industry.

4.1. INTRODUCTION

In the today's world more advance technology has introduced. So, today in the real world with the help of augmented reality (AR), which is a new technology, computer graphics (CG) can be overlayed. The best feature of the advancement of technology is that it describes the whole structure, all the problems and in a summarized way. Staring point is provided by the chapter to anyone who is interested in researching and using AR.

MR is a term which refers to a multi-axis spectrum of areas that cover virtual reality (VR), AR, telepresence, and all other related technologies. Allowing user to enter and interact with the artificial environment is done with the computer-generated 3D environment, which is a term used for VR. In the computers artificial world, users can immerse themselves to varying degrees which may either be a simulation of some form of reality or the simulation of a complex phenomenon.

The main purpose of telepresence is to provide problem solving abilities to a remote environment and to extend operator's sensory-motor facilities. So, it means sufficient information is received by the human operator about the teleoperator and task environment, as telepresence is defined as a human-machine. The information which is received is displayed in a natural way because with this operator will feel the physical presence on the remote site.

Telepresence and VR are very much similar, as in VR, the aim is to achieve the illusion of presence within a computer simulation, whereas in telepresence, the aim is to achieve the illusion of presence at a remote location.

Between VR and telepresence AR can be considered a technology. It has been seen that in VR the environment is completely artificial and in telepresence it is completely real, in AR the user sees the real world augmented with virtual objects.

Three aspects must be kept in mind while designing an AR system:

- Combination of real and virtual worlds;
- Interactivity in real-time; and
- Registration in 3D.

The augmented scene can be shown with the help of wearable devices, like Head-Mounted Displays (HMD), but other technologies are also available.

> Markets expect the Augmented Reality market figure to reach $61.39 billion USD by 2023. Augmented reality will soon out perform its counterpart markets. It's all huge numbers from here on. By the year 2023, the AR market could be $61.4 billion USD!

Besides the above mentioned three aspects, there is one more which could be incorporated: Portability. It has been seen in all virtual environment (VE) systems that due to the device limitations, the user is not allowed to go around much. However, a user really walks through a large environment is something which is needed by the AR application. Thus, the issue of portability becomes important.

There is a registration by the name 3D, and for such applications, it becomes more complex. Unregistered, text/graphics information using a monocular HMD is generally provided by the wearable computing applications.

These systems are not based on AR system by narrow definition but these are the system which are based on the see around setup. Therefore, for more general applications, wearable display devices and computing platforms that are used in AR are developed.

The growth and progress of AR is remarkable, since the field of AR has existed. So, after analyzing the rapid growth in the field of AR, many conferences specialized in this area were started. Apart from the conferences international workshop and symposium on AR, the international symposium on MR, and the designing AR Environments workshop has also been started.

For several years, AR is considered as a hot topic in software development circles, but with the release of products like Google Glass, it started getting the renewed focus and attention.

To augment sound, video, graphics, and other sensor-based inputs on real world objects using the camera of the device, a technology that works on computer vision based recognition algorithms is AR. To make virtual elements to be the part of the real world than it is a good way to reduce the real-world information and present it in an interactive way.

The overlayed information is displayed by the AR in the field of view (FOV), and further, this can take anyone into a new world where the real and virtual worlds are tightly coupled. The display of AR is not just limited to desktop or mobile devices. The perfect example for this is google glass, which is a wearable computer with an optical head-mounted display.

When the user captures the image of the real-world object than there is simple AR use case which detects a marker, which triggers it to add virtual object on top of the real-world image and displays on your camera screen.

4.2. COMPONENTS OF AR

In recent years AR get into the limelight with the help of some applications like Snapchat, harry potter, Pokémon GO. With the advancement of technology, the way of interaction with the world has become more enhanced, and the place looks more vibrant to live in (Figure 4.1).

Figure 4.1: Components of AR.

Source: Image by unsplash.com.

Many people think that AR is the new technology, but people will be shocked and amazed to know that this technology was invented in bac 1800s.

AR is defined as the enhanced version of reality by the Merriam-Webster dictionary. To overlay digital information onto live camera feed AR is created by technology. In other words, it is defined as a technology that allows content to look like the part of the real world. As in VR users are transported completely into a completely digital world, so AR is different from AR.

Hardware, software, and the application are the three components that every AR system comprises of. With the help of taking example of the smartphone, this concept can be explained.

4.2.1. Hardware

Through which the virtual images are being projected is known as hardware equipment. So, in this case, equipment is considered as the smartphone. There is a requirement of sensors and processors that can support the high demands of AR, for AR to work on these devices. Here are some of the key hardware components:

- ***Processor:*** The main role of the processor is to determine the speed of the phone, as it is the brain of the device. This is done to see whether the device can manage the heavy AR requirements or not, apart from its normal phone functions.
- ***Graphic Processing Unit (GPU):*** Visual rendering of the phone's display is handled by the GPU. High-performance GPUs is required by the AR to create the digital content.
- ***Sensors:*** Device which gives the ability to make sense of its environment.

Common sensors required for AR include:

- **Depth Sensor:** To measure depth and distance.
- **Gyroscope:** To detect the angle and position of your phone.
- **Proximity Sensor:** To measure how close and far something is.
- **Accelerometer:** To detect changes in velocity, movement, and rotation.
- **Light Sensor:** To measure light intensity and brightness.

These above-mentioned specifications are crucial for AR to function properly on the devices. This is considered as one of the big reasons that why only the later generations of mobile phones have AR capabilities.

To acquire and display the data and information, and process it is the main feature of the hardware components.

- **Input-Sensors:** From the real environment there are different types of sensors that respond to physical or chemical stimuli and it also provides some important data for the development of the system (Table 4.1).

Table 4.1. Input Hardware for AR

Hard-ware	System	Type	Example
Input	Sensor	Optical	Cameras, infrared
		Magnetic	Compasses
		Inertial	Accelerometer, gyroscope
		Others	GPS, depth

- **Output-Display:** Into wearable and non-wearable the devices for displaying the information can be divided. But it can also be classified into optical, video, and projection devices (Table 4.2).

Table 4.2. Output Hardware for AR

Hard-ware	System	Type	Examples
Output	Wearable	Optical	helmets, glasses
		Video	Head-up display
	Non-wearable		Smartphones, tablets, PC
		Projection	Projectors

- **Optical See-through Display:** To mix with the real scene optically the virtual contents are projected onto the interface. For example, DAQRI, HoloLens, Moverio.
- **Video See-through Display:** It has two work modalities. One works with camera and screen in handheld devices, and one uses HMD devices.

4.2.2. Software

The AR magic begins with the software, which is the second component of the AR system. The main example of AR software is AR Kit (Apple) and AE Core (Android). There are three fundamental technologies consisted by these programs that allow them to build AR experiences.

1. ***Environment Understanding:*** To map its surroundings out, this allows the user's phone to detect prominent feature points and flat surfaces. After doing this, the system can then place virtual objects accurately on these surfaces.
2. ***Motion Tracking:*** In relation to the environment, this lets your phone to determine its position. After this virtual object can then be planted on their designated spots on the image.
3. ***Light Estimation:*** This gives your phone the ability to perceive the environment's current lighting condition. Virtual objects can then be placed in the same lighting conditions to enhance realism.

Notice how the hardware and software work hand in hand. If the hardware does not possess the sensors to measure light intensity, the software's light estimation capabilities will be of no use.

To interpret the acquired data to transform and augmented it, there are some main characteristic of software components that are:

- **High-Level:** There are many tools which help in the development of applications, known as SDK (software development kit), such as Vuforia, ARKit, ARCore, Wikitude, ARtoolKit, EasyAR, LayAR.
- **Low-Level:** It consists of different areas which may vary according to the application or the developer's needs, such as programming libraries, Computer vision, CG, Image processing, HCI.

4.2.3. Application

The last component of an AR system is the application itself. Software should allow the AR applications to run on the smartphone, but it does not possess the AR features. 3D objects and filters are some of the AR features that come from the mobile application themselves.

IKEA, Pokémon GO and Snapchat are the applications which have their own triggering logic and own database of virtual images. These applications help to map them out onto the live images and also pull virtual images from their database.

4.2.4. Working Principles of AR

Marker-based tracking and Marker-less tracking are generally two ways that applications trigger. Marker-based tracking: To trigger the AR features, this mode requires optical makers such as QR codes. The application recognizes it and superimposes the digital image on the screen by pointing your mobile camera at one of these barcodes.

Marker-less tracking: In the case of Snapchat, the real-world object is the face of the user, and this mode is based on the object recognition. On the basis of marker less tracking the AR applications works, and they get triggered when they recognize certain real-world features.

4.2.5. AR Implementations in the Marketing World

It is said that in the marketing world, the applications for AR are plenty. It is like a dream that in the world there will be a time where pointing a camera at a brick-and-mortar restaurant shows the reviews of the food there.

Along with this providing people with special discount codes when they use AR lenses to look at the shop. To find the specific information, this method is used, as it is the faster method and it is also a more interactive means of engaging the customers. AR is considered as the most powerful tool when it is used wisely.

4.3. AR DEVICES

The latest and most promising term in technology is AR. Though, the concept goes back to the time of 1900s when Howard Grubb patented the collimating reflector. AR is not bounded only to a headset and can being tested anywhere from phone to wearable and even on projectors. The case is different in AR than the VR.

It is said that AR has the biggest potential for mass consumption, though VR and MR also fall under the same extended reality spectrum.

- ***Types of AR:*** Regarding the use cases, there are four types of AR devices available right now as follows:
  - Heads up displays (HUD);
  - Holo-graphics displays;

- Smartglasses; and
- Handheld.

While using AR, no dedicated hardware is required, and this the best thing about AR and can utilize on phones, tablets, and computers. As per the increasing demand in the mobile phone, the day will come soon when phones, AR glasses, and headset will coexist together (Figure 4.2).

Figure 4.2: An illustration of augmented reality device.

Source: Image by unsplash.com.

Some of the AR devices that will change the future are discussed in subsections.

4.3.1. Microsoft Holo-Lens 2

HoloLens is developed by Microsoft and it is by far the closest glass to a perfect AR device. It has the capability of integrating 3D objects better than any of the available AR devices, as it has two 2k 120 Hz display. To interact naturally with the virtual world HoloLens 2 tracks eye and hand movement.

With the minimizing and adjusting the size, HoloLens 2 can be considered as the fearless leader of AR.

4.3.2. Magic Leap One

In 2018 there was a most trending start-up by the name Magic Leap which raises about 3 billion USD from multiple giant companies. By the end of 2018, a revolutionary AR device has been launched by the name Magic Leap One.

The Magic Leap One enables the user to see the AR images in the real world. Nowadays where technology dominated the whole world, if any device does not remove the user from the real world, then that device is considered as the best option.

4.3.3. Epson Moverio

Epson Moverio AR devices allow you to see a virtual display floating alongside the real world, unlike HoloLens and Magic Leap One. This display is related to the eyes and the head of the user instead of the world. This makes sure that the display will stay positioned towards the user all the time.

4.3.4. Google Glass Enterprise Edition

At the time of introduction of the google glass, it faced the failure. Along with the features of recording video and capturing images without even notifying people around made the device unpleasant to users.

However, by using Google's artificial intelligence (AI), and features like customizing glass as needed, it is now a perfect fit for doctors, mechanics, or entrepreneurs, and others alike Google found a way to recover from it. While staying connected with the virtual world to make things more productive and efficient, it allows them to use both hands to work.

4.3.5. Vuzix Blade AR

Few years back Vuzix Blade AR took the place of the Google Glass. However, instead of an awkward independent display in the corner it has an integrated display in the middle of the glass. To do the basic operations, it uses Alex, and it has another feature that it can also be paired with smartphones. Moreover, the Vuzix Blade AR camera has a red indicator to let everyone know if the camera is on, unlike the Google Glass's camera.

4.4. APPLICATION OF AR

Different aspects of the AR experience are controlled by the computer program. The AR application is referred as the computer program. A very simple example of an AR application is a simple AR browser. With the help of AR browser, it will be easier to make it appear like a specific 3D object is placed on a specific fiducial marker placed in the real world.

Between the content and an AR application it is important to draw the difference used within the application. The same AR application can be used in many different contexts when this distinction is executed skillfully.

For example, with the purpose of teaching an art appreciation course, an AR application that places simulated sculptures in a real-world space could be repurposed by using different content and using it to train soldiers regarding the kinds of landmarks they need to locate when they are deployed to a place they have not yet experienced in the real world.

With the help of the AR application, it is possible to interact with the various sensors, devices, and displays used in the experience. Many of these lower-level tasks are handled in AR libraries that are used by many different AR applications in the actual practice.

Adding the virtual value to every industry from retail to industrial manufacturing is a promise that holds by the AR apps, headsets, and the smart glasses. To solve some of the biggest problems and pain points, AR is already showing potential. There is no need to wait for AR, for more years to make a big impact across the board.

Some of the best use cases for AR technology that are set to emerge soon, from education to remote work are discussed in subsections.

4.4.1. Medical Training

In many areas AR tech holds the potential to boost the depth and effectiveness of medical training and from operating MRI equipment to performing complex surgeries. Anatomy will now get learned by the students at the Cleveland Clinic at Case Western Reserve University. And they learn this by utilizing an AR headset, as it allows them to delve into the human body in an interactive 3D format (Figure 4.3).

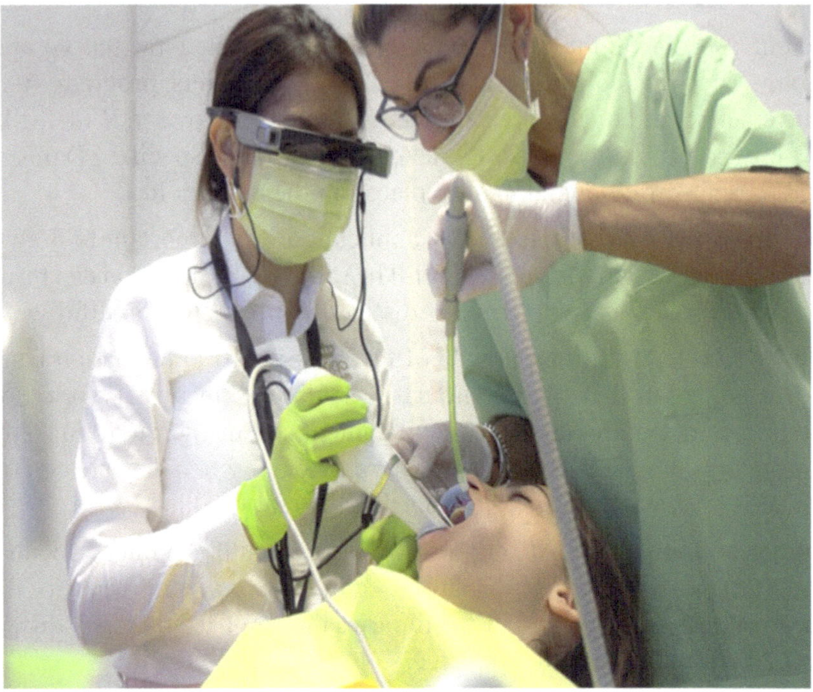

Figure 4.3: Application of augmented reality in medical science.

Source: Image by unsplash.com.

4.4.2. Retail

In the today's market smartphones are used by the retailers in the retail environment, to compare the prices or lookup for the addition information on products they are looking for. There is renowned brand Harley Davidson which is famous for its motorcycle.

This brand has developed an AR app that shoppers can use in-store. This is considered as the one of the great instances of a brand making. With the help of this application, user can view the different colors and features of the bike and can also customize it accordingly.

4.4.3. Repair and Maintenance

For repair and maintenance of the complex equipment, the AR is used. The staff has begun to make the use of the AR headsets and glasses for repair and maintenance of car motor or an MRI machine. And with the help of AR, they can get useful information on the spot, it also suggests potential fixes and point out the potential trouble areas. This will get stronger from machine to machine as the technology grows and it can also feed the information to AR headsets directly.

4.4.4. Design and Modeling

In the case of interior designing and construction, during the creative process, AR is helping the professionals to visualize their final product. With the use of headset architects, engineers, and design professionals get an opportunity to see how their design might look like and, they can make virtual on the spot changes.

Using AR headset visualization, urban planners can even model how entire city layouts might look. Any job which is related to the design or modeling jobs that involve spatial relationships and they are the perfect use case for AR tech (Figure 4.4).

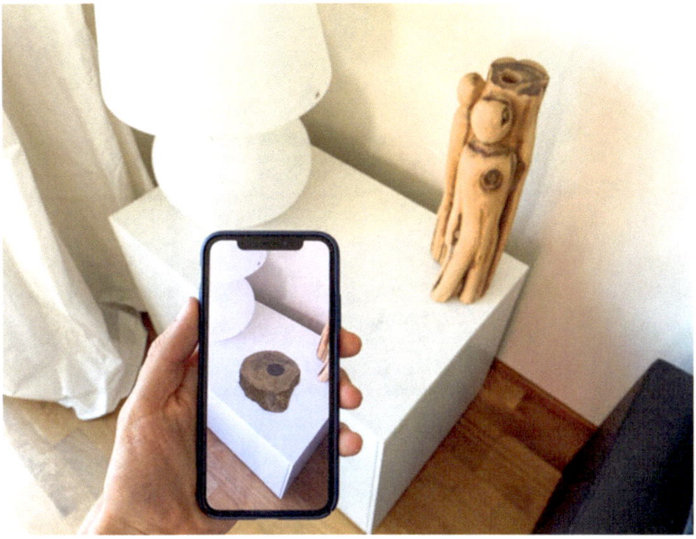

Figure 4.4: Application of augmented reality in design and modeling.

Source: Image by unsplash.com.

4.4.5. Business Logistics

Across many areas of business logistics where there is a need to increase efficiency and cost savings, AR presents a variety of opportunities.

Transportation, warehousing, and route-optimization is included in this. In some of the warehouses of the shipping company like DHL has already implemented smart AR glasses, where lenses display to workers the shortest route within a warehouse to locate and pick a certain item that needs to be shipping. In today's business environment providing workers with more efficient ways to go about their job is one of the best ROI use cases.

4.4.6. Tourism Industry

In recent years technology has gone a long way towards advancing the tourism industry. It has been reviewed from various sites like TripAdvisor and some informative sites like lonely planet. But for travel bands and agents, AR presents a huge opportunity to give potential tourists an even more immersive experience before they travel (Figure 4.5).

Figure 4.5: Application of augmented reality in tourism industry.

Source: Image by unsplash.com.

With the help of AR vacations and selling tris become a lot easier in the future. As it gives the opportunity to take a virtual walkabout for any place.

4.4.7. Classroom Education

In many schools and classrooms, technology plays a very important role, as teachers and educators are now ramping up student's learning experience with AR. For example, there is an application by the name Aurasma which is already being used in the classrooms so that students can view their classes over the smartphone or tablet for a good learning environment.

A full map of the solar system can be viewed by the students who are learning astronomy, or the students attending music class might be able to see the musical notes in real-time while learning to play an instrument.

4.4.8. Field Service

Field service technicians are required in every area or field, whether it is a small air conditioner or it is large as wind turbine. So, with the help of AR headset or glasses technician can view whatever they are repairing and can also diagnose the problem and can fix it also. It is the faster way of resolving the problem instead of solving it manually.

4.4.9. Entertainment Properties

Building strong relationship is very important in every industry, and in the entertainment industry, it is all about building good relations with the audience and your brand characters. Viewers of the harry potter are always hungry for more additional content (Figure 4.6).

Figure 4.6: Application of augmented reality in entertainment properties (mobile application: Pokémon go).

Source: Image by Unsplash.

Viewers also feel that they know the characters of harry potter. As a great marketing opportunity, AR is seen by the entertainment brands to build the bond between characters and their audience.

4.4.10. Public Safety

Smartphone is used by people for every small to big thing, like in case of any emergency or something related to that. Immediately people start browsing their smartphone to know what is going on, where to go. A promise is shown by the AR to figure in solving public safety puzzle.

For example, if there is a situation of earthquake or the fire, than it will help the responders to figure out who needs help and the best way to get the safety. Responder wearing AR glasses can be alerted to danger areas and show in real-time individuals that need assistance while enabling to still be aware of their surroundings.

4.4.11. Content

Content is considered as the key to any AR application. All the objects, ideas, stories, sensory stimuli, and "laws of nature" for the experience is included in the content. It also consists of all aspect of AR application that do not fall into any of the other ingredient categories. What actions take place during the experiences the law of nature govern. This may include computational simulations, game rules, or any other aspects of the content that are under computer control.

4.4.12. Interaction

In one or another way every AR experience must be interactive. This is considered as the most typical way in the world, as it allows the participant to view/perceive the world from different physical points of view. Apart from this participant may interact with the experience by pressing a button, by making physical gestures, by speaking commands, or by any number of different actions (Figure 4.7).

Figure 4.7: Application of augmented reality in the form of glasses.

Source: Image by Unsplash.

4.4.13. Technology

Technology is involved by every AR experience. Much more sophisticated technology is required than others, but all have a base level of technology. Information about the real world is to be gathered at a minimum AR experience with some sort of sensor.

With the real world, some form of computation is needed to integrate the virtual elements of the experience. Along with this, some mechanism to display the virtual elements of the experience is required.

4.4.14. The Physical World

In the physical world, every AR experiences takes place. The physical-world is considered as the key part of the AR experience. In some cases, a generic space is used to represent the physical world at large. In some cases, the experience must take place where the experience indicates.

For example, like a Christmas tree, an AR experience is to decorate them (real) Eiffel Tower. Then virtually a user can experience and explore the different ways of decorating it.

4.5. AUGMENTED REALITY (AR) VS. VIRTUAL REALITY (VR): WHAT IS THE DIFFERENCE?

The AR and VR considered as technology twins, and they are often confused. Below the explanation of VR and AR in plain terms and along with this some AR and VR cases.

In the same breath, AR and VR are often talked about, and this is done because to measure the market for these related capabilities. The managing director of augmented world says that there are many commonalities between AR and VR, and in the future, these technologies will be the two side of the same coin. Further, it allows anyone to experience the AR or VR on the same device depend upon the requirement of the real world.

Though, today AR and VR are two different things but VR get the more coverage as compared to AR, as AR can be more nebulous given the various ways in which it can be deployed.

AR and VR requires business leader to plan different strategies for implementing these technologies in their organizations, since AR and VR are not one and the same. So, for more clarity, let us study the difference between AR and VR with some real-world examples.

The important differences between AR vs. VR (Table 4.3).

Table 4.3. AR vs. VR

Augmented Reality (AR)	**Virtual Reality (VR)**
• The system augments the real-world scene. • User always have a sense of presence in the real world in AR. • AR is 25% virtual and 75% real. • This technology partially immerses the user into the action. • 100 Mbps bandwidth is required by AR. • There is no need for AR headset. • End-users are still in touch with AR, with the real world while interacting with virtual objects nearer to them. • • Both real and virtual worlds is enhanced by AR.	• Completely immersive virtual environment. • Visual senses are under control of the system in VR. • VR is 75% virtual and 25% real. • This technology fully immerses the user into the action. • At least 50 Mbps connection is required by VR. • Some VR headset device is needed. • By using VR technology, VR user is isolated from the real world and immerses himself in a completely fictional world. • • For the gaming world, it is used to enhance fictional reality.

4.5.1. AR vs. VR: Key Differences

User is allowed to experience the real world with the help of AR, which has been digitally augmented or enhanced in some way. On the other hand, VR removes the user from that real-world experience, replacing it with a completely simulated one.

Because complete involvement is required by VR and VR devices shut out the physical world completely. There are some AR capabilities which are delivered by lens on the smart glasses that deliver, and apart from this they are transparent. In determining the best use cases for each, understanding these differences is critical.

4.5.2. AR vs. VR Use Cases

In the real world, there are many cases in which users need to be connected with each other, and for that, AR applications are best suited at a high level. Remote assistance, on-the-job training, remote collaboration, and computer-assisted tasks are some of the AR enterprise solution.

After the result of research which was conducted of both technologies, it was found that AR to be well suited for industrial use cases, mainly for workforce training and product maintenance. With the help of AR tools, there are some companies who are facing knowledge gaps and expertise loss as workers retire are capturing that knowledge digitally and sharing it with less-experienced workers.

One example: Usually, company use to write out training documents, but with the help of AR professionals use to record and narrate their task and after that it is used for workers to train in a digitally assisted way. The company says workers trained in this fashion tend to retain 80% of what they have learned, compared to 20 to 30% when they read a manual.

For simulation or complete immersion VR applications are best suited. Think remote collaboration with 3D elements, point-of-view training, and virtual tours. To improve training for orthopedic surgeons and nurses, Johnson & Johnson Institute developed VR software. VR is use by many big companies like Walmart to create both unlikely scenarios (such as weather emergencies) and common ones to give associates a first-hand training experience without disrupting operations.

4.5.3. AR vs. VR in the Future

It is said that AR and VR will continue to be applied in different ways during the next three to five years. Different value propositions and different purposes is served by them. However, if wearables become more mainstream in the enterprise, these capabilities may converge more with time.

4.6. AUGMENTED REALITY (AR) VS. VIRTUAL REALITY (VR)

The user can directly participate with one another in a physical environment through AR. This is done through virtual objects, which are computerized imitation implanted in the environment. This visual environment is facilitated in VR for users completely.

This allows us to explore the idea of dualistic interaction between the mind and the body. This allows the user mind to explore the VE and its effects with response to one another.

The idea of presence belongs to the psychological state of being present while experiencing meditated and unmediated environment. Where on one hand the user cannot accept the reality of the environment in VR, on other hand the user can sense and except his/her environment in AR. Furthermore, the user does not accept the existence in VR meditated perception but in AR they accept the unmediated reality and its existence.

OHMD's (optical head mounted displays) consist of various sensors such as IMU's (inertial measurement units) that consists of accelerometer, gyroscope, and magnetometer which is used by AR and VR systems. Moreover, it also consists of various microphones, and a camera to display virtual objects, environments, etc. Through computerized environment OHMD's can convince the user of being in the mediated and conscious environment.

4.7. EXAMPLES OF OHMD'S

4.7.1. Microsoft Holo-Lens

On January 21st, 2015 at the event of windows, Microsoft came up with the most advance technological hollo lens, which amazed the whole world. It is a headset with transparent lenses and with a shine presented at Redmond, Washington Headquarters.

This headset is such that once put by a user, the whole surrounding is being changed. One can see the 3D objects floating around the living room or a home on the walls, one can see the virtual characters and can see the subject in the midair. It allows the user to interact with high-definition holograms which are one of a kind and fully unbound holographic OHMD which one can feel in the surrounding (Figure 4.8).

Figure 4.8: An example of overhead mounted display.

Source: Image by unsplash.com.

This new technology of HoloLens provides user with new advance experience, which is combined with AR, VR, and live videos. Through hand gesture gaze and voice the customer relates to the environment. It completely modifies and enhances the real world unlike replacing it with a virtual one.

4.7.2. Oculus Rift

VR headset is developed and manufactured by the Fb Inc. which is a division of Oculus VR is known as Oculus Rift. On March 28, 2016, it was released. The display of the oculus rift is OLED (organic light-emitting diode). To provide 3D audio effects, it has integrated headphones. Along with this, it also has rotational and positional tracking sensors (Figure 4.9).

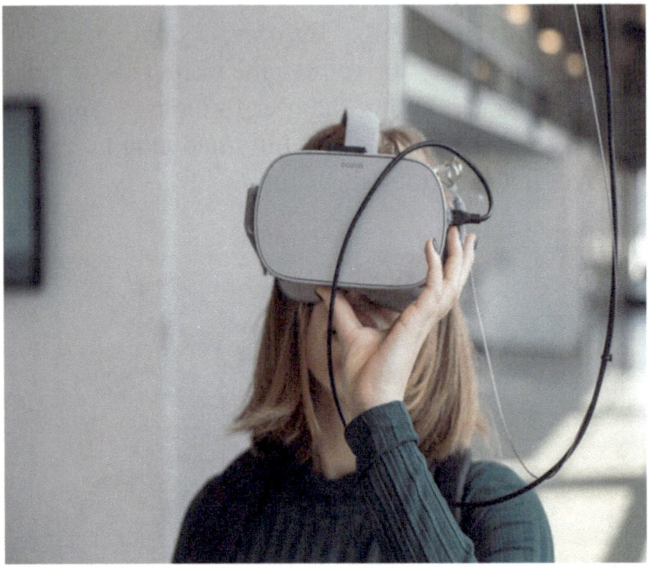

Figure 4.9: Picture depicting overhead mounted display of 'Oculus Rift.'

Source: Image by unsplash.com.

While walking, sitting, and standing in the same room, user can use the HMD because Oculus Rift creates a 3D environment.

With the aim of having the full attention of the user into the virtual world, Oculus Rift block the view of the outside world. Once the user relates to HMD, the screen in front of his eyes makes him isolate from the world outside, and he is unable to sense the real world, and it makes the user more distant as well. Moreover, Oculus Rift is only meant to be a seated experience which can be only done at one place.

4.8. ADVANTAGES OF AUGMENTED REALITY (AR)

With each passing year, the technology of AR seems to be increasing day-by-day and everyone wonders if this would be the breakthrough year for AR. From AR/VR previous projects of 2019, there has been seen an increase of 78.5% in the spending for the year 2020. According to IDC reports $18.8 billion has been the spending budget for 2020.

Moreover, it is expected that major spending for this budget would be done in commercial sector. There are some advantages of AR that are mentioned in subsections.

4.8.1. Augmented Reality (AR) Takes the Store to the Customer

Major furniture companies like IKEA have shown interest in the technology of AR. In order to provide their customer with visual experience on how their homes or offices would look like with their furniture, IKEA has been taking utmost advantage of this technology. They can give their customers this exclusive experience with this new age technology of AR by letting them step into the virtual world.

4.8.2. AR Enables Unique Opportunities for Immersive Reality

AR technology has been seen to be explored in various fields such as social media. Many content creators and writers are rooting in to find various ways to connect AR and for AR-based social network, which could be another set mark for the future.

Applications like Snap chat is again using the same technology of AR by enabling its users to generate location-based filters in their own cities through their users. Moreover, this can be easily used in the mobile devices, and one can also send and receive location-based messages through this application.

4.8.3. AR Provides Children with New Levels of Interactive Experiences

More companies such as BIC kids have enhanced the experience of AR by providing Children with DrawyBook application. This allows them to do their artwork through tablet and create and do their artwork with multiple features. They can change colors, shapes, and can also view their drawing on the application.

4.8.4. AR Provides Safety Technology for Automobiles

In the area of Automobiles companies like Jaguar is providing its users with enhanced technology such as Windscreen which provide the driver with actual data where user can see any preceding objects coming through the windshield. It comes as a go in advance technology for the awareness of the driver.

With the coming year of 2020, more companies like GMC are also involving with the technology of AR. With the advanced technology of AR, the GMC is introducing "transparent trailer view." This uses a camera where they can place over the image to have the actual view for the trailer being towed. This can be seen on the dashboard screen.

4.8.5. AR Allows Workforce Training Without Risk

Now as days the AR technology is also being developed and used in the workforce area. With the country like United States, their Army has been contracted with Microsoft to have the license with HoloLens technology worth $480 million.

This technology is expected to be used for various purposes such as training, simulation, and combat. Furthermore, this automation includes various features such as night vision to be used in training as well as in actual combat. With this "Integrated Visual Augmentation system" soldiers would be able to enhance their skills of effectiveness and can also help them in becoming a combating machine.

4.9. ADVANTAGES OF VIRTUAL REALITY (VR)

4.9.1. Training

For various training purposes in different sectors, VR provides a whole new learning experience. This is mostly been used in precarious environment such as in the field of Health. Various doctors make use of VR technology to avoid unnecessary medical accidents while doing operations.

In the airline industry, Pilots use VR to get trained in landing a plane; moreover, it also comes as a tool for firefighters to prep before the first fire occurs. In the coming future, various advantages of VR could be seen in various sectors as alluring learning tool all over the world.

4.9.2. Conferencing

VR for Business could become a new norm. With applications like Facebook live or Skype combined with VR could provide us with the real time experience where one can be present in the same room with another person instead of just seeing the person on screen.

One can work together just like in real meeting and conferences even though sitting miles. One can easily feel the other person just like in real word moreover this could be predicted as a future for conferencing world.

4.9.3. Convenience

Organizations and companies can be benefitted from VR to save a lot of time and money for its workers in terms of traveling. VR helps in creating and exploring various sectors such as virtual marketplace. On one hand, where shoppers can try any garment virtual on the other hand, they can also look how their home will look with that Arabian rug. Furthermore, with people like Architects who must travel around in order to evaluate or finalizing designs, they can do the same virtually. This comes as great financial saving.

4.10. DISADVANTAGE OF AUGMENTED REALITY (AR)

There have been various issues and concerns regarding the AR, whether it will bend the real world or not. It depends on a lot of factors such as data distribution in computerized formed as well as user generated information. These all factors are problematic considering the following subsections.

4.10.1. Issues about Privacy

There have been many problematic issues regarding the privacy rights and security in the usage of AR. As it involves different types of data in its formation, such as collection and analysis through big data, it causes a threat to various legal concerns. Virtual recording and analyzing the surroundings and the environment in real-time can raise some serious concerns regarding one's privacy.

Moreover, user information like their biometric data, their usage history is also been collected through AR systems. There is strict need for developers to follow data protection laws such as GDPR in order to be more severe with the usage of these systems.

4.10.2. Dangers of Reality Modification

When the real and the digital worlds are mixed up, there comes the possibility of inability to distinguish between the both. This involves a lot of risk taking and some serious accusations. Such as in case of gaming world where games like 'Pokémon' it leads to lot of accidents and deaths for its involvement with environment world. Moreover, for users it becomes difficult to distinguish in his consciousness between physical world and digital world and dangers it could involve.

There is a need for developers to take care of the digital elements being involved in the creation. There should be necessary standards to be followed in order to educate users to not be dominated by AR and forgetting about the real world outside and which could also result in dangerous issues.

4.10.3. Implementation Requirements

Implementation of new technology is never an easy process as not everyone has the capability to do so. As, it required a lot of resources by the business to implement this technology. Though, small firms always face disadvantage because of the shortage of resources. Implementation of AR system is very costly.

Capable processing capabilities is required to run the AR application smoothly. As, it should be noted that new technologies and models are required by AR. The main criteria for the development of AR are the development of AI technology. There are some specific forms such as machine learning, natural language processing, and computer vision, among others.

4.11. DISADVANTAGES OF VIRTUAL REALITY (VR)

4.11.1. Lacks Flexibility

In the classroom, a student has the full flexibility of asking the question and can provide any suggestion. But, in VR it is not possible, as in the case of VR headset anyone can make use of the same program in all the session and along with this there is no scope for positive interaction.

4.11.2. Ineffective Human Connections

There are many sets of disadvantages which comes along with the VR. As the predictable education system is based on the level of individual human communication and on interpersonal connections, so the concept of VR is totally different, and it is only about the software which is used by the student.

4.11.3. Getting Addicted

It is said that in VR addiction is extremely common. To the virtual world, students can get connected very easily. As, it has been seen that some section of the population is getting addicted to video games and the rest. One can even get addicted to harmful drugs in the world of VR.

4.12. WHERE IS AUGMENTED REALITY (AR) USED?

Here, are an important applications of AR technology:

- Text, images, and videos are embedded by the AR apps;
- To display digital content on top of real-world magazines printing and advertising industries are using AR technology apps;
- AR technology allows you for the development of translation apps that helps you to interpret the text in other languages for you;
- With the help of the Unity 3D Engine tool, AR is being used to develop real-time 3D Games (Figure 4.10).

Figure 4.10: Image depicting an overhead mounted display.

Source: Image by unsplash.com.

4.13. WHERE IS VIRTUAL REALITY (VR) USED?

Here, are important applications of VR:

- VR technology is used to build and enhance a fictional reality for the gaming world.
- VR can use by the military for flight simulations, battlefield simulations, etc.
- VR is used as a digital training device in many sports and to help to measure a sports person's performance and analyze their techniques.
- It is also becoming a primary method for treating post-traumatic stress.
- Using VR devices such as Google Cardboard, HTC Vive, Oculus Rift, or users can be transported into real-world and imagined environments like squawking penguin colony or even the back of a dragon.
- VR technology offers a safe environment for patients to encounter things they fear.
- Medical students use VR to practice and procedures.
- Virtual patients are used to helps students to develop skills that can later be applied in the real world.

4.14. HOW AR AND VR WORK TOGETHER?

One cannot say that AR and VR only works separately in order to give the user an experience with new dimension, these technologies are being combine. This provides user with the experience of the fiction world where a person gets engaging experience both with real and virtual world. It completely brings the user to an amazing virtual world.

4.15. WHAT'S HOT IN AUGMENTED REALITY (AR)?

At build 2015, the show was takeover by Microsoft when it launches HoloLens. It provided with the expansion in the world of AR when the HoloLens was used to create ocean wave, it was a revolutionary picture being painted by AR through HoloLens.

Currently, Microsoft is working with holograms in order to bridge an interactive opening between and one's living room. With the help of HoloLens, a user can surround himself with the windows app. It is said that to infiltrate audience home from a marketer's view, this becomes one more, intensely immersive, and promising way.

To receive the development edition of the HoloLens Cramer was fortunate to be one of the first agencies and the experiential future for our clients is already looking brighter. With the help of AR technology applications for product demonstration has started building.

In the year 2016, it has been seen that in the form of Pokémon Go. AR has taken the center stage. The viral sensation that got Pikachu and Charizard out of the Gameboy and onto your front lawn, whether you wanted them there or not! This was the first major example of AR finding mass market acceptance and infiltrating our daily lives.

To expand the sensory experiences in 2017, VR and AR has started making dramatic leaps forward as startups and it find the ways to introduce smell and touch to expand your sensory experiences. With the use of haptic feedback technology company Immersion has introduced Touch Sense Force to bring player's hand into VR worlds. There is a lab by the name Stanford University's Virtual Human Interaction in which researchers must resist eating foam doughnuts as they experiment with adding scent to VR.

Also, there are many design and engineering companies which are the likes of the solid works and are demonstrating their commitment to immersive design with AR and VR related partnerships, including NVIDIA, Microsoft, Lenovo, and HTC Vive beyond the obvious media and entertainment applications for AR/VR technologies.

4.16. THE STATE OF THE AR/VR ADOPTION RATE

In the today's market AR and VR considered as more relevant and they are gaining speed as well but for the small marketers and tech enthusiasts they are considered as more than anything a toy for a small minority.

In real-time to render 3D environment, both are hindered by the ability is the main reason behind this. AR is less important because in AR, a user is just adding onto it as the environment already exists, but there is still a problem with creating high resolution, life-like objects, persists.

This can be equated with the early video games. For example, Nintendo N64 and for and at that time goldeneye was a remarkable game and still it has a major fan base count in the billions, but with low polygon count. The simplest form of 3D is polygon and polygons that make up an image is with the higher 3D resolution.

Now a day's trailers of the new games are like the trailer of the movies and games are getting better day by day as they are having polygons. So, this promises well for the future of AR and VR.

4.17. CONCLUSION

AR and VR attain high standards in many fields like entertainment, science, medicine, visualization, and annotation, robot path planning, military aircraft, etc., and it also allow experience that are evolving more commonly than expected. AR is said to be ahead of VR, as there are several products already available in the market, but there are some limitations with VR.

It blocks the user's interaction with the surroundings, apart from providing a whole immersive experience. It is said that AR devices are successfully commercialized as they do not completely disconnect people from the real world.

AR is ahead of VR, as there are several products already in the market. With the VR-induced sickness and nausea, VR OHMD's also have been associated, which can be considered a problem for some people.

While attending another task AR headset can remain productive, and it also does not require any user to stand at one place. This reason is considered as important because of apart from being behind VR on the current development curve, and on the enterprise market, AR is expected to have a bigger impact.

REFERENCES

1. 3Pillar Global, (2013). *Augmented Reality-Introduction and its Real-World Uses*. [Online] Available at: https://www.3pillarglobal.com/insights/blog-posts/augmented-reality-introduction-and-its-real-world-uses/ (accessed on 29 April 2021).
2. Chen, R., (n.d.). *Your 3-Minute Guide to Augmented Reality (AR): How Does it Work?* [Online] Constructdigital.com. Available at: https://www.constructdigital.com/insight/how-does-augmented-reality-ar-work#:~:text=Every%20AR%20system%20comprises%203,smartphones%20to%20expound%20the%20concept.&text=The%20hardware%20refers%20to%20your,case%2C%20they%20are%20your%20smartphones (accessed on 29 April 2021).
3. Clark, S., (2019). *Top 5 Benefits of Using AR for Your Business*. [Online] CMSWire.com. Available at: https://www.cmswire.com/digital-experience/top-5-benefits-of-using-ar-for-your-business/ (accessed on 29 April 2021).
4. Cramer, (n.d.). *What is the Difference Between AR and VR? Cramer*. [Online] Available at: https://www.cramer.com/insights/the-difference-between-ar-and-vr/?__cf_chl_captcha_tk__=a4dd55b2056d2a9d638ddfc23f915d0d384aebba-1614792751-0-AXZ4WqioQcK17GzpeFzPCM9nmBqPRJPMp_lx5C-vi8ra29Uu2XIc47lg723SkI8z-tLcy8qpI9RDATveBLwUqhSQGdh5hmLu3rve67JLAqFUwTaWHdM5BTZJ8WULAkjrY-ukHGlHhDPZgSAlfH4ZeaVEfLfLNgEFIv_HY3IMup2ysHRyuTVQp261dRHAhAgFwB2i_twQQBekhbvJHJJ_Np_-TQtxN8DWUSu5wYhq3udRrcJN4raCBMdylLsy9KJokNJsP67YmJwj5U4cwO-9X7BU9sWhj3N5Dp08CpVNqMJ_STkwF0i1qo3QC8yCBHq1YDK7D-L9enNDz52Yt_zwjMnohHO2-QC6FtXs9yeiXqG_Uq_05DFNe4UI-IVq7XKLeyd7fncuaj0jzmKTm2ox_ZDU-5tc8ufJ12pMOzLwpRoA-ZrXwcfDHRYuKfXUzQDg2yL34cfBQUka5CDNEbhfVJKGrY5bmX3B3Zmb8bB6CCih1ZzsRfsuEPlae0hwFNWp08z7j89bda85Svb0izSZbhJoPysjRUi6Aeq8NYCce1-lPTmO9Bt15yTIYXxl35kAa_FhxVnqm-ZXCPWJBQdXo74 (accessed on 29 April 2021).
5. Enterprisersproject.com. (2009). *Augmented Reality (AR) vs. Virtual Reality (VR): What's the Difference?* [Online] Available at: https://enterprisersproject.com/article/2019/10/ar-vs-vr-whats-difference (accessed on 29 April 2021).

6. Guru99.com. (n.d.). *AR vs. VR: Difference Between Augmented Reality, Virtual Reality*. [Online] Available at: https://www.guru99.com/difference-between-ar-vr.html (accessed on 29 April 2021).
7. Inc.com. (2021). *10 Real Use Cases for Augmented Reality*. [Online] Available at: https://www.inc.com/james-paine/10-real-use-cases-for-augmented-reality.html (accessed on 29 April 2021).
8. Ivankov, A., (2019). *Advantages and Disadvantages of Augmented Reality | Profolus*. [Online] Profolus. Available at: https://www.profolus.com/topics/advantages-and-disadvantages-of-augmented-reality/ (accessed on 29 April 2021).
9. LiveTiles, (2017). *Three Benefits of Virtual Reality | LiveTiles.* [Online] Available at: https://livetilesglobal.com/3-benefits-virtual-reality/#:~:text=VR%20can%20save%20organizations%20time,alone%20is%20a%20monetary%20godsend (accessed on 29 April 2021).
10. Medium, (2020). *AR 101-Components of the Augmented Reality System (Part 3).* [Online] Available at: https://medium.com/@mafda_/ar-101-components-of-the-augmented-reality-system-part-3-878c71e68069 (accessed on 29 April 2021).
11. PaleBlue, (2021). *Best AR Devices for 2020 | PaleBlue.* [Online] Available at: https://pale.blue/2020/02/03/best-ar-devices-for-2020/ (accessed on 29 April 2021).
12. Sciencedirect.com. (n.d.). *Augmented Reality System: An Overview |* ScienceDirect Topics. [Online] Available at: https://www.sciencedirect.com/topics/computer-science/augmented-reality-system (accessed on 29 April 2021).
13. Technologies, B., (2018). *The Benefits and Disadvantages of Virtual Reality*. [Online] YourStory.com. Available at: https://yourstory.com/mystory/c6ba370f23-the-benefits-and-disad (accessed on 29 April 2021).

CHAPTER 5

Human Being in Virtual Environment

CONTENTS

Understanding the human being role in the in the virtual environment (VE) is very crucial. In the chapter, human being in VE, the impact of virtual reality (VR) technology on human lives is explained. The chapter also tries to give the answer of whether we are living in VR. It also shed some light on the human performance efficiency in virtual worlds. It also demonstrated gesture interaction in VR. It also discusses how does VR impact the human condition? At the end, the chapter explains the good and the bad of escaping to VR.

5.1. INTRODUCTION

Virtual reality (VR) has the potential to create new realities where people can be what they prefer or choose to be, instead of what they are told to be or born to be. A lot many people have a perception in mind that VR will eventually make them isolated.

It would not be correct always that a person who has been very much passionate about virtual worlds for the past several years, since before the technology was capable of supporting VR. It is something that will help people understand as well as relate to each other, in a way that could be hard to imagine.

VR can take shared experiences to new levels of intimacy. A person can put himself in the experience of another person. VR can create a living, breathing virtual world, it aids people in choosing their own professions and build their own community.

They can accumulate power, trust along with wealth and lose it all in a matter of minutes, just like in the real world. In several ways, life in the virtual world feels just as meaningful as the real life. The friendships a person make, the betrayals he experiences, and all the other emotions he feels are just as authentic.

5.2. THE IMPACT OF VIRTUAL REALITY (VR) TECHNOLOGY ON HUMAN LIVES

After the development of internet, VR appears to be the next big innovation. The reason being, it has virtually changed the way businesses are done with its various applications. These days, every top app development companies are trying new and effective ways of applying the VR technology with their apps (Figure 5.1).

Figure 5.1: VR influencing human lives.

Source: Image by NewGenApps.

5.2.1. Pilot Training

Pilot training is the oldest application of VR. Before a new pilot is allowed to fly a plane in real, he has to be tested first on simulation. If he passes the simulation test, then only he is allowed to fly a real plane. Due to this reason, every aviation school has its VR app development company. The simulation is designed in such a way that it looks very real, and a practitioner will feel like he is flying a real plane.

5.2.2. Medical Application

These days, surgeons make effective use of virtual artificial to check through a body of a patient, so that he can accurately determine the location of tumors and how they can reach them. At present, every big healthcare provider is working with a VR app development company in order to develop medical VR apps.

For instance, an experienced VR app development company developed an effective app for VR therapy. This app has been proved to be really useful in the field of medical science. It can further be used for people that have some kind of phobias. Through this app, people are made to face their respective phobias head-on so that they can somehow get over it and start living a normal life.

In addition, the VR therapy can also be used to treat people having certain kind of mental issues by developing soothing and peaceful environments as well as simulations. For example, a lot of soldiers and marines have to deal with PTSD when they get back from the battlefields.

VR therapies can be used in an effective manner to treat them. Through therapies, they will be exposed to different types of battlefield scenarios to help them get over the mental stress and start to lead a normal life.

There are a lot of medical VR-based apps that are being utilized for teaching student surgeons how to carry out operations. Actually, each and every medical VR-based app development company knows how to design such apps that will eventually stimulate an operation. Thus, a student can only be allowed to deal with a real operation once he is able to perform well in the simulation.

5.2.3. Watching of Movies

These days, many cinemas distribute VR headsets to their viewers before watching a movie. When movies are viewed through those headsets, they seem so real as well as engaging. In order to make the whole movie immersive, some of the cinemas even install hefty speakers in distinct parts of the hall.

Therefore, with the speakers and the view provided by the headsets, collectively provide a great immersive experience. Basically, it makes a person feel like he too is a part of the movie. When viewed through the VR platform, the scary movies appear much scarier. The suspense and anxiety also get doubled while watching movies through such platforms.

Well, addiction is a little challenge with watching a movie through those headsets. Once a person gets used to watching movies through headsets. He will be having issues watching movies on the conventional tv set, they will appear boring and not engaging, as VR makes people participate in the movie. This is the reason why there are two types of cinemas now, the ones that have adopted the technology of VR while the other that are being run out of business slowly.

5.2.4. In the Automobile World

Almost every automobile manufacturing company has its own app development company based on VR. With the advent of VR technology, 3D images of the prototype of a proposed car can be seen and evaluated by all concerned automobile engineers before the manufacturing of a physical car.

This technology basically allows them to assess the engine, the exterior and even the interior so they can make their suggestions before the car is manufactured. This is cheaper as well as convenient than having to manufacture a prototype before corrections are made.

For example, Ford is said to be the first automobile manufacturing company to use this VR technology. Then, Toyota also followed suit and various other manufacturing companies have joined the bandwagon.

5.2.5. In Military Training

VR is also being applied to military training. This is convenient and much cheaper, and it also makes learning much more interesting as well as entertaining. Various battlefield scenarios can be simulated on VR platforms, and the soldiers are further trained on how to handle such situation.

The training has now become more effective as every battlefield scenario can be simulated. Over the past few years, soldiers that were trained with the applications of VR have been compared with their counterparts that got physical training. It was also found out that both categories of soldiers have been performing equally.

5.2.6. In Video Games

Video games are very interesting when played through a VR platform. It becomes more immersive as well as engaging, and most of all it becomes more addictive and compelling. The games simply playing on the device will never provide as much fun as playing it on VR platforms (Figure 5.2).

Figure 5.2: VR in gaming.

Source: Image by Pinterest.

This is why many people still prefer going to game houses when they have more than five games on their device. Apart from video games, there are various ways by which VR can be utilized for the entertainment purpose. Universal studios provide unlimited fun by their virtual tour rides. Some of the scenes in the ride need VR headsets, while many others do not.

5.2.7. In Museums

Some of the museums have uploaded their collection of items on various platforms of VR. All a person requires is to have both VR headset as well as access to any of the platforms. He or she will be capable of accessing the museums from the comfort of home at any time of the day. It can be said that this is nothing but a real convenience.

Even if people are in any other continent, they can easily access the museum. Imagine how much time as well as money a person will save by accessing the museum from headset based on VR. Well, at present just a few of the museums have uploaded their collections on the VR platforms, while various others are working on uploading theirs.

For example, British Museum in London recently uploaded its collections and also, it launched a virtual tour of the museum. On the other hand, the American Museum of Natural History has also uploaded its collections.

5.3. DO WE LIVE IN VIRTUAL REALITY (VR)?

There is an idea that people are living their lives inside of a digital simulation, a kind of huge computer program, similar to the size of the universe. The simulation at first started a long time back, sometime in ancient past developing VR that people see today in their daily lives.

This idea has a major impact on the core of most intimate beliefs almost about everything such as, from disease and healing, to peace and war, life, and death, angels, God, religion, UFO's and many more. Simulation may sound like a science fiction movie.

5.3.1. What Is the Simulation?

So, simulation can be defined as an experience of the environment in a comparatively say way, reducing the risks of injuries to people. It is clear that the idea of digital environment is not something new today, as people have already used simulated environments in various places or situations. At present also, people are making use of simulations in many ways in daily life.

5.3.2. The Characteristics of a Simulation

- The world of simulation has a place where it starts and where it ends. Whatever happens in between is completely based on mathematical algorithms of cycles in such patterns that are repeated again and again on different scales.
- There are certain rules that direct the simulation as well as the idea that people in the simulation world become much more familiar with the environment so that they can learn those rules and life gets better.
- In the simulation world, the user always has access to an external reality. They can easily tap into the guidance if they have or face any kind of issue, or if they need any kind of clarification, they can always communicate beyond the simulation itself.
- In the simulation, the user cannot feel or experience the difference between simulation and the real world.

5.4. HUMAN PERFORMANCE EFFICIENCY IN VIRTUAL WORLDS

Computer speed and functionality, synthetic sound, image processing, and tracking mechanisms have been combined together in order to offer realistic virtual worlds to people (Figure 5.3).

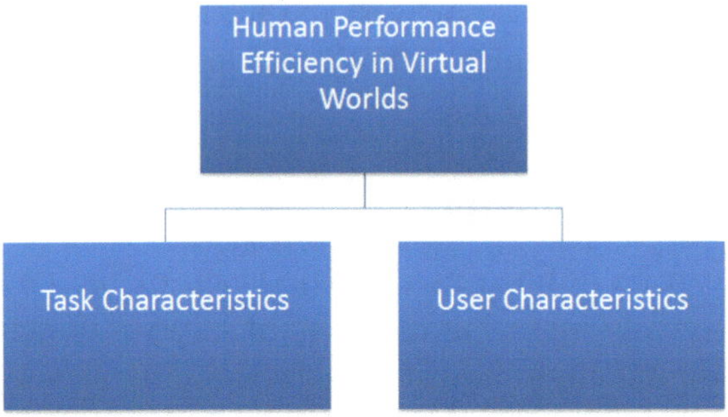

Figure 5.3: Human performance efficiency in virtual worlds.

An important advance still needed for VEs to be effective is to define how to extend the efficiency and effectiveness of human task performance in virtual worlds. In several scenarios, the task will be to attain and comprehend information being represented in the VE.

> As Wann and Mon-Williams (1996, p. 845) stated, in such cases, "the goal is to build (virtual) environments that minimize the learning required to operate within them, but maximize the information yield."

Maximizing the overall efficiency of the information transferred in VE will need developing a group of guiding design principles that enable efficient and intuitive interaction so that users can freely access and understand data.

The design tactic that are generally used for more conventional simulation systems (e.g., teleoperated systems) was usually "trial and error" because of the paucity of general, non-system-specific design standards (Burdea and Coiffet, 1994).

If VE tasks are designed in such an ad hoc manner, although, the probable advantages of the virtual world, such as transfer of training and enhanced understanding, may be compromised.

While it is almost challenging to interpret the significance of several human-factors issues needing consideration, it is certain that if humans cannot efficiently perform in VEs, thereby compromising the overall effectiveness of the transfer of training or the HVEI, then further pursuit of this technology may be unproductive.

In a way to evaluate the overall effectiveness of a VE, a means of evaluating human performance efficiency in virtual worlds is needed firstly. This can be said easier than to do it actually. In contrast to past HCI studies, VE performance measures are required to emphasize on more than task outcome to be effectual.

Because of the multifaceted nature of HVEI, it may be imperative to form effective multi-criteria measures in order to interpret human performance in VEs. Factors that contributed the most to human performance in VEs probably comprises of the navigational complexity of the VE, the degree of presence offered by the virtual world, and the performance of its users on benchmark tests.

If individuals cannot traverse effectively in VEs, then their potential to engage in needed tasks will be sternly confined. Wann and Mon-Williams (1996) have discussed the object and edge information and data, as well as the dynamic global changes in textures and features that are essential to attain navigational performance effectively.

Beyond defining such design principles that stipulate the crucial cues to upkeep navigation, a designer must find how their user would be able to move in a virtual world. While some developers have incorporated "various per-mutations of windowing to let people to move freely" about VEs (Laurel, 1994), newer designs such as spirals, portals, sliders, and two planes hold promise for improved VE interaction.

Nilan (1992) has searched the mapping of a user's cognitive space to the design of the VE to improve the overall navigational performance. Darken and Sibert (1996) have developed modalities and mediators to support with VE navigation. These methods are likely to attain a varying range of accomplishments prior to concentrating on standard(s) for VE navigation.

Because of the recent issue of users becoming lost in greatly less-complex, hierarchical computer environments (Sellen and Nicol, 1990), the development and design of standardized navigational techniques that would guide individuals in upholding spatial orientation within VEs may prove to be one of the most critical concern to resolve in VE design (Darken and Sibert, 1996).

Without proper means of moving about VEs, developers would not be able to enhance the human performance. One possible approach to tackling this issue is to cultivate a means of gauging the "navigational complexity" of a VE (i.e., how challenging it is to move from one point to another point).

Such a measure would likely be reliant on the spatial mediators and sensorial cues (e.g., maps) provided by an environment, as well as on the navigator spatial ability. Navigational difficulty varies mainly between present VE and VE designs, thus possibly becoming a diagnostic tool for assessing the effectiveness of a given VE design (for instance, redesign could be focused on areas that have greater navigational difficulty).

Wayfinding (Vishton and Cutting, 1995), Mental maps (Siegel, 1981), dead reckoning (Gallistel, 1990), spatial orientation (Adolfson and Berghage, 1974; Thorndyke and Statz, 1980), homing (Adler, 1971), time to collision (Lee, 1976), vestibular functions (Howard, 1984), and geographical orientation (Clark and Malone, 1952), are some of the fields with technical knowledge with respect to the perceptual issues embraced with navigation in VEs.

5.4.1. Task Characteristics

One of the most critical tasks that will unswervingly impact how effectively humans can perform in virtual worlds is the tasks nature that are being performed. Some tasks may be suited exclusively to virtual representation, while others may not be executed effectively in such environments.

To validate the practical usage of VE technology for a given task, in comparison to substitute tactics, the use of a VE would require to enhance the task performance when transported to the real-world task because the VE system exploits on a primary and individually human sensory, information-processing, perceptual, or cognitive capability.

It is highly imperative to evaluate the sorts of tasks for which these categories of advantages can be attained by using VE technology. Wann and Mon-Williams (1996) have recommended that data and feature analyzes (e.g., visual detail enlargement, dimensional assessment, contingency evaluation, design envisioning) and data visualization are feasible tasks for extracting such advantages.

Viable tasks can be stipulated systematically by attaining an understanding of the connection between task characteristics and the corresponding VE features (i.e., real-time interactivity, stereoscopic 3-D visualization, multisensory feedback, immersion) that efficiently assist their performance both within the VE and upon conveyed to the actual world scenarios.

Some indication of the advantages of stereoscopic visualization comes from work by Tendick, Kim, and Stark (1993). In examining the advantages of stereo cues in a virtual pick-and-place task, it was witnessed that stereoscopic display was far superior in comparison to a monoscopic display.

Remarkably, though, functioning with the monoscopic display was just as efficient as the stereoscopic display when advancement in visual perspective display were provided (i.e., projection lines and ground grids). Comparable advantages from stereo were validated by Merwin, Wickens, and Lin (1994).

Smith, Pepper, and Cole (1981) have also instituted that stereoscopic display performance is usually high-class in comparison to monoscopic performance; although, they institute the range of enhancement was dependent on the visibility, task, and learning factors.

Kim et al. (1993) implies that, for simple telemanipulation tasks, cognitive cues, and monocular depth cues (amassed through past experiences and learning) may be sufficient for effective performance.

It was argued that when tasks become more intricate, and the cognitive and monocular cues provided are lacking or inadequate, stereoscopic cues will augment the overall performance. These suggestions may elucidate why Hancock, Kozak, Chrysler, and Arthur (1993) found no noteworthy transfer-of-training advantages from VR training, in comparison to actual world and no training conditions, on a pick-and-place task.

5.4.2. User Characteristics

An imperative aspect impelling human VE performance is the consequence of user disparities. These disparities, to provide an easy analysis, can be with the throughput (e.g., perceptual or cognitive styles), input (e.g., interpupillary distance), or output (i.e., human performance).

Noteworthy individual performance distinctions have already been noticed in primary VE studies (Lampton et al., 1994). User characteristics that greatly impact VR experiences need to be recognized in a way to design VE systems that assist these exclusive wants of users.

User disparities have already been stated to impact the sense of presence (Weghorst and Barfield, 1993) and cybersickness felt by VE subjects (Parker and Harm, 1992). What is not as understandable are which user types have a delicate influence and which considerably amend virtual experiences.

In order to identify which user personalities may be dominant in VEs, one can scrutinize studies in human-computer interaction (HCI). While both HVEI and HCI comprises of HCI, because of differentiations between them (i.e., HVEI is normally from an egocentric perspective.

While HCI is usually from an exocentric frame of reference), earlier HCI studies may not unswervingly employ to HVEI but can offer probable visions. A noteworthy amount of research has emphasized on comprehending the effects of user characteristics in HCI.

Even though imperative, this research has yet to offer steady indication as to which aptitudes are the most persuasive in HCI. The lack of definitive theories and user models makes puzzling the process of recognizing user characteristics important to VE interaction.

Yet, while it might be calm to say that these "impedance matches" may be job exclusive, it is highly believed that there are enough simplifications possible such that dependence on the "task-specific" justification should be evaded at all costs, since this remark is so non-informative.

5.5. GESTURE INTERACTION IN VIRTUAL REALITY (VR) (FIGURE 5.4)

Figure 5.4: Gesture interaction in VR.

Source: Image by iStock.

5.5.1. Human-Computer Interaction (HCI)

Human-computer interaction (HCI) refers to the research in the design, application, and execution of computer systems by human. It is a kind of technology that aids studying humans, computers, and the interaction between human and computers.

With the advent of computing devices and some related technologies, the interaction between humans and computers has become a key part of daily life such as work, communication, shopping, etc.

The graphical user interface based on keyboard and mouse is the most commonly used user interface. Although, the user interface based on the window, icon, menu, pointer (WIMP) interface paradigm is restricted to 2D (two-dimensional) planar objects, which cannot interact with the objects present in the 3D (three-dimensional) space, decreasing the naturalness of the interaction.

The interaction between humans is done by communicating through a variety of different aspects such as language expression, emotional interactions, and body movements. The new generation of HCI must have different features like human-centered, multimodal, and intelligent. As a result, it is important to propose a new interface paradigm, create a new user interface, and adopt a new interaction method.

In real life, people make use of their hands to perform different operations like move, grasp, and rotate. In the process of communication, people naturally attract the attention of other people just by their hand movements. Gesture is the way by which people express their will under the stimulus of consciousness.

Hence, gesture interaction has two functions now. One of which is the hand movement, while another is the specific meaning of the gesture. A study showed that compared with other parts of the body, the hand being a common communication device is highly suitable for the application of HCI. Gesture interaction has key features-natural, intuitive, and flexible.

Thus, it is important for some of the users with physical disabilities like hearing impairment and visual impairment, to interact via gestures.

5.5.2. Gesture Interaction Technology in Virtual Reality (VR)

HCI being an important supporting technology of VR, offers a large variety of interactive modes based on distinctive purposes and functions, enabling people to acquire immersive feelings in 3D virtual scenarios.

VR, as an emerging technology is a brand-new interactive HCI interface. Consequently, a new generation interface paradigm post-WIMP and non-WIMP has been defined, and also, the interaction extends from a mouse- and keyboard-based 2D graphical user interface to a natural user interface or supernatural user interface.

Different from traditional WIMP, in a natural user interface or a supernatural user interface scenario, the user interacts in a 3D environment just by simulation of VR or interactive scenario design, which eradicates waste of resources produced by repetitive manufacturing and controls the cost of production.

With the sustained development of interactive technology, VR has been largely used in various fields, for example, game entertainment, education, and medical care. In the future, the interaction could be like the movie 'Player One' in that users could capture the movement of people through wearable devices, becoming avatars, and interact with other people living in the virtual world.

As an increasing number of scholars and companies invest in the research of gesture interaction technology and its related applications, gesture interactions have involved a large part of the field of natural interaction.

In order to put on daily life operations in a VE, it is important to track as well as identify the hand movements, to make the computer understand the true intention expressed by gestures, and also to perform related tasks. Simultaneously, to make people have a natural interactive experience in the process of VR, a relevant feedback should be given to the user.

Gesture can be conscious or unconscious movement of the hand, arm, or limb. The gestures in the process of interaction can be differentiated as per different spatiotemporal operation behaviors, different modes of interaction, different semantics, and different ranges of interaction.

In the early scenarios of interaction, the procurement of gesture signals was completely based on wearable sensor devices like data gloves; with the innovation of mobile devices such as mobile phones and tablets, touch screens became really popular amongst all; as a result, the procurement of gesture signals has developed into visual signal acquisition from computer cameras.

In the past few years, acquisition of the information based on high-tech equipment such as electromyographic signal acquisition has progressively become a research focus. Due to the variability as well as complexity of the gesture, how to utilize the current technology to process the input signal received by the device, how to find out the spatial position as well as posture of the hand, and how to acquire a correct result of recognition, have a huge effect on the impact of consequent gesture interactions.

The technology of gesture recognition is primarily categorized into gesture recognition based on touch devices, gesture recognition based on wearable sensor devices, and gesture recognition based on computer vision. With the advent of gesture recognition technology, it has been applied to various aspects of VR.

5.5.3. Gesture Definition and Gesture Classification

Gesture is a movement or posture of the upper limbs of the user. Using gestures, people can effectively express their intentions f interaction and thus, send out corresponding interactive information. In general, gestures conveyor send intentions or information via physical movements of the limbs, face, or body.

In the VR environment, the movement or posture of the users' body is utilized as input, while the posture information of the user through the interaction device in the physical world is used as the system input. In the process of interaction, users cannot just produce gestures using, finger, palm, and hands, but also, they can produce gestures as an extension of the body using different tools like pen, mouse, and glove.

Users can efficiently send simple commands to the system using gestures such as selecting, moving, and deleting. In addition, they can also express more complex intentions like controlling virtual objects, switching the current interactive scene, and performing virtual actions.

In some other conditions, similar gestures may have distinctive meanings because of certain factors like geography or culture. The basic understanding of gesture semantics is highly related to the cultural background. The gestures, therefore can be summarized and classified according to different situations.

5.5.4. Classification Based on Spatiotemporal Status

In general, gestures are classified into static gestures and dynamic gestures, according to the spatiotemporal operating state. A static gesture is defined as a static spatial posture of a finger, palm, or arm at a definite moment. Usually, it represents one interactive command and does not include time-series information.

On the other hand, a dynamic gesture is defined as the changes in the finger, palm, or hand over a specific period of time; also, it contains time information.

The key difference between the two is that static gestures only comprise spatial gestures without causing alterations in the spatial position, while on the other hand, dynamic gestures are characterized by the movements in spatial locations over a period of time, like waving gestures, that are one of the typical dynamic gestures.

In addition, the dynamic gestures also comprise conscious dynamic gestures as well as unconscious dynamic gestures, in which the unconscious movements during physical activity are known as unconscious dynamic gestures, while gestures for the purpose of communication are known as conscious dynamic gestures.

5.5.5. Classification Based on Gesture Semantics

As per the cognitive and behavioral psychology, Pavlocvic et al. categorized gestures into intentional movements and unintentional movements. Intentional movements are considered as the gestures, on the other hand, unintentional movements are referred to as the movement of arm/hand without conveying any meaningful information.

Further, gestures are divided into manipulative and communicative gestures (based on whether they contain the characters of communication or not). Manipulative gestures can be regarded as the action of control of the state of the objects by using arm/hand movement, like moving and rotating different objects by the movement of the arm.

Whereas, the communicative gestures can be regarded as the customized gestures with a specific information function. In a situation of natural interaction, they are generally accompanied by the verbal communication, especially comprising action gestures and symbol gestures.

Action gestures consist of mimetic gestures that mimics the actual interaction behavior and a deictic gesture with a direction-indicating function. While symbol gestures have a fixed-constraint meaning and can be further be divided into referential gestures and modalizing gestures.

In continuation with conscious and unconscious gestures, the conscious gestures are sub-divided into:

- Emblem gestures;
- Regulator gestures;

- Affect display gestures;
- Illustrator gestures.

Emblem gestures are also known as quoted gestures. It is basically a direct conversion of short language communication and the gist are related to the cultural background, like waving hand which means goodbye.

Regulator gestures control the progress of an interaction. Affect display gestures are the ones that express emotions or the intent of a user, but less relevant to the cultural contexts. Illustrator gestures are a description of the user's language communication state, highlighting the main parts of the expression, and these are highly related to the cognition as well as language skills of a communicator.

Further, McNeil categorized the illustrated gestures into five different categories:

- Beat gestures;
- Iconic gestures;
- Deictic gestures;
- Metaphoric gestures; and
- And cohesive gestures.

Beat gestures are generally a repetitive action, which have the features of short-time and fast speed. Iconic gestures describe the action or image through the movement of the hand. The deictic gestures represent the real position or the abstract position of the target. Metaphoric gestures define abstract things by using simple but meaningful expressions. Cohesive gestures transfer interactive tasks.

Unconscious gestures, that are also known as adaptor gestures, primarily refer to unconscious communication habits generated during communication.

5.5.6. Classification Based on Interaction Mode

As per different interaction methods, gestures can be categorized into media gestures, direct contact gestures, and non-direct contact gestures based on the mode of interaction. Media gestures are a single-point input device via physical contact, like a mouse, trackball, or a joystick, in order to transmit the acquired data series into the computing system.

Direct contact gestures can be described as the gesture operations directly in the input device by part of the body or a physical object tool. When a user touches the device, for example, screen, then the gesture interaction is turned on, and the touch movement is considered as an intermediate state of interaction. Direct contact gestures also comprise a single touch gesture as well as a multi-touch gesture.

The single-touch gesture is defined as a gesture operation provided by the user by a single point of interest using a pointing input device such as pen or mouse, or a body part such as a hand. A pen gesture is a distinctive single-touch gesture that records the trajectory movement of a pen and utilizes a logo or symbol for intention expression.

As compared to a single touch gesture, a multi-touch gesture is a gesture operation over input devices and multiple regions of interest of the body. Non-direct contact gestures denote to a kind of operation of conducting an interactive command via a body part or a physical object without being in physical contact with the system.

Different from the direct contact gestures, the non-direct contact gestures do not touch any input device or surface during the operation directly, for instance, vision camera-based gesture interaction.

5.5.7. Classification Based on the Scope of Interaction

As per the various scope of interaction, gestures can be categorized into "stroke" gestures and "mid-air" gestures. The former one is particularly used on a support surface based on a handwriting gesture of the upper part of the wrist.

It means, the user uses the tool or hand directly in order to move on the surface of support and communicate with the computer for the exchange of information, such as the touch screen interactions. The mid-air gesture is completely based on the limbs of the user or interactive devices moving in a space without a surface of support.

Meaning that the limbs of the user or the interactive tools are not in contact with the computer system while interacting in a 3D space. In a relative manner, the stroke gesture has a simple model of interaction and low flexibility while the mid-air gesture has a higher degree of freedom of interaction, and thus more modes of representation.

Compared with the former one, the range of interaction of the latter is a bit larger, however, it lacks the ability of capturing for action details, and consequently, the recognition accuracy of the detailed action is not sufficient.

As per the different modes of display, mid-air gesture interaction senses can further be categorized into 3D virtual senses based on a display of a traditional desktop and VR scenes based on a helmet-like device. However, the traditional desktop display can show a 3D virtual scene, the user generally requires to do the gesture interaction within a definite distance of the fixed display.

Display movements that are based on the helmets or some other off-the-table environments ultimately permits the user to stay free from the restrictions or limitations of the device as well as the environment. Further, this increases the range of gesture interaction.

Certainly, while using the helmet display, the spatial range is amplified for the activity of users, the user cannot interact on a fixed surface. In such cases, it could be an efficient choice to make use of non-contact 3D gestures for interaction.

5.6. HOW DOES VIRTUAL REALITY (VR) IMPACT THE HUMAN CONDITION?

Naturally, human beings crave for connecting with each other. At present, VR can provide improved realities that can eventually aid in fostering such relationships, mainly for people who are isolated. However, the level to which VR contributes to deeper connections depends mainly on how the technology is being utilized (Figure 5.5).

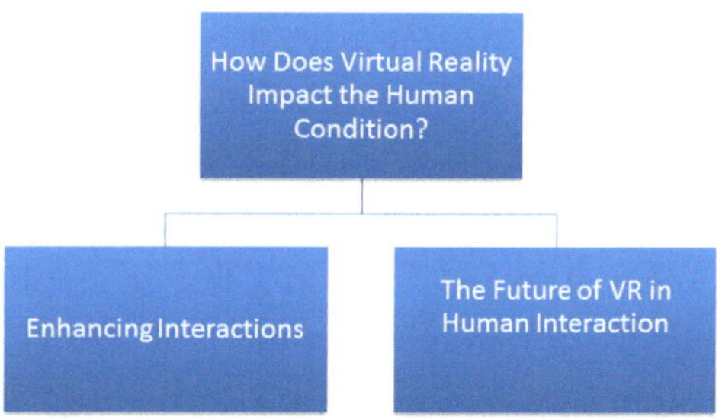

Figure 5.5: Role of VR in impacting conditions.

A survey "technology and the human condition" done in year 2017 found that mobile technologies believe that VR has the potential for improving the human connection. In total, survey respondents felt a bit optimistic.

Out of all, 63% felt VR improves rather than taking away from the human experience, while on the other hand, 57% specifically feel that VR will help them to develop empathy for each other.

5.6.1. Enhancing Interactions

VR creates modern ways of interacting with each other. Such technology can remove the limitations or barriers to interacting and communicating, specifically for people who are restricted by physical, emotional, and psychological issues.

For people who are mostly homebound, and unwilling or unable to leave the house, VR can foster a sense of familiarity, belonging, and personalization. It actually allows those people to connect to one another even when they are restricted by physical geography.

According to Aaron frank, principal faculty at singularity university, a Silicon Valley, "If simply hearing someone's voice and seeing a few body-tracked mannerisms is enough to convince my brain that I'm hanging out with someone I know, I can only imagine future social VR experiences that build on that sensation to make me believe I'm actually hanging out with friends."

It can be said that VR has the ability to help people see the world from a different point of view, which can eventually provide them a better understanding of people who are different from each other.

5.6.2. The Future of VR in Human Interaction

When VR is compared with augmented reality (AR), it is found that their capabilities of reaching mainstream market success are really different. In the industry, most of the technologies believe that AR will attain market success much sooner than VR, however, there is little consensus about when.

However, many people believe that AR is not of much use to the general public. It has the potential to be useful in a few years, but it requires to be more mature as well as affordable as people feel cost is a major barrier for mainstreaming consumption. Certainly, this technology is valuable to some industries or professional environments.

It can be said that the expense of VR is not yet justified providing a lack of effective consumer value that technology offers. At present, edge cases of VR have proven to be very profitable, however, industries need to find an efficient way to make the technology more accessible, predictable, and affordable too, in order to become global with mainstream consumers.

5.7. THE GOOD AND THE BAD OF ESCAPING TO VIRTUAL REALITY (VR)

In Silicon Valley, in the year 1985, a ragtag band of programmers started exploring the notion of VR from a tiny cottage in Palo Alto. A team headed by a 24-year-old Jaron Lanier aided in making VR a buzzword in the mid-to-late 80s and got significant investment, prior to filing for bankruptcy at the decade's end.

In spite of mass media interest from publications such as Scientific American and Wired, the technology was not there, or it was highly expensive—and the audience was a tad too niche. Save for some fruits of its early research, purchased in sum by Sun Microsystems, VPL's sole legacy has been its popularization of the term "virtual reality."

Since then, around 35 years have passed, and the landscape has eventually shifted in the favor of VR. For example, Microsoft developed a project named 'HoloLens,' a kind of headset that creates holograms of high-definition.

It is thick and black lenses that make use of an advanced depth camera, sensors along with different processing units to process a large number of bouncing light particles, so as to project the holographic models on the kitchen counter, or to take the wearer on a hyper-realistic to Mars.

Considering other instances in the field of virtual/AR, Google has invested $542 million in the augmented-reality startup Magic Leap, while Sony as well as Samsung are both developing virtual-reality headsets. Much was made of Facebook's $2 billion purchase of VR Kickstarter darling Oculus Rift, as Mark Zuckerberg made it very clear that the company was playing the long game: "*One day, we believe this kind of immersive, augmented reality will become a part of daily life for billions of people*."

All these signs point out to a future filled with VR. All such examples show that the potential application of VR is beyond count. A person could have breakfast at the louver beside the winged victory of Samothrace, followed by a lunchtime spelunk through Thailand's water caves.

There are highly immersive video games and different movies based on VR. In addition, Barcelona's BeAnotherLab has created an empathy application for the oculus rift that ultimately allows its users to swap genders.

If, in the future, VR becomes a part of day-to-day lives of people, then a greater number of people would prefer spending majority of their time in virtual spaces. The futurist Ray Kurzweil predicted that, "by the 2030s, VR will be totally realistic and compelling and we will spend most of our time in VEs… We will all become virtual humans."

Theoretically, such escapism is nothing novel-as the critics of increased TV, internet, and smartphone applications will say, but as the technology of VR sustains to blossom, the world that they will generate will eventually become highly realistic, and create a high potential for overuse.

Such technological paradigm shift brings an extent of immersion, different from all of the others that has come earlier, and the handwringing has already started. Keeping early doomsday predictions aside, here the question arises, can virtual escapism can ever be used for good?

The oldest documented research on escapism dates back to the 40s and 50s reportedly, when researchers first started examining the link between life satisfaction and media consumption or usage. In the year 1996, Peter Vorderer, a professor at the University of Mannheim, attempted to define the term 'escapism,' and wrote, "*escapism means that most people have, due to unsatisfying life circumstances, again, and again cause to 'leave' the reality in which they live in a cognitive and emotional way.*"

As people cannot actually leave reality, so the concept of escapism seems to lack precision. From this, it can be said that VR is a game-changer. With VR, it is really possible that in spite of just escaping reality by concentrating on a TV show (for example), people may prefer to replace an unhappy reality with a happy and better VR.

The idea of a life that is lived completely online, or outside of the usual society, is highly considered as unhealthy as well as dangerous. There are several reports of self-imposed social isolation that exemplify the negative side of withdrawal.

In Japan, since the 1990s, the term "*hikikomori*" has been used in order to describe the estimated 500,000 to 1 million Japanese citizens who refused to leave their homes. As per Dr. Takahiro Kato, a psychiatrist working at a *hikikomori* support center in Fukuoka, Japan, a lot many *hikikomori* show obsessive as well as depressive tendencies, while on the other hand, a minority of them seems addicted to the Internet.

In addition, there are the infamous World of Warcraft players who lose themselves in their huge online universe. In the year 2004, Zhang Xiaoyi, a 13-year-old from China, allegedly committed suicide after playing WoW for 36 consecutive hours, so as to *"join the heroes of the game he worshipped."*

On the other hand, in the year 2009, a girl of three-year age from New Mexico passed away tragically from dehydration and malnutrition. Her mother, after the death of her kid said that she used to spent 15 hours a day playing the game.

To 'The Guardian' in 2011, a former Warcraft player Ryan van Cleave explained that living inside the World of Warcraft appeared preferable to the drudgery of day-to-day life, when he played 60 hours a week. Groups such as WOWaholics Anonymous have been created in order to help the former players who became highly invested in the game.

However, these were the extreme instances that share a common root with lesser forms of negative escapism. Another definition of unhealthy escapism could be like-escapism is the effect on the essential fabric of living (in the context of family, friends, and social commitments). This is connected to Abraham Maslow's hierarchy of needs, which ranks love and a sense of belonging just after basic physiological and safety needs.

Critics often point to how screen-saturation has affected the way people fulfill those needs (in a negative way), while others have further explored how VR might only aggravate the issue.

Ignoring the fact that future applications of VR also include the power of connecting with real human beings all around the world, this is a new platform of communication, and it is possible to find love and relationships online.

According to Jim Blascovich and Jeremy Bailensen, "The Internet and virtual realities easily satisfy such social needs and drives—sometimes [they are] so satisfying that addicted users will withdraw physically from society."

Blascovich, a psychology professor at the University of California, Santa Barbara, and Bailensen, of Stanford University's Virtual Human Interaction Lab, observed the penalties of a VR-centric future, noting that as the platforms of virtual-reality become affordable and mainstream, the pull of spending more time in VR may prove hard to resist.

The proliferation of an affordable VR system will largely increase the size of the population for whom very perceptual as well as psychological experiences are available. Although, such immersive escapes are not necessarily a bad thing.

A virtual second life can easily replace the real-life of few people, but this can either be good for them or bad. There is no clear instance from which it can be concluded that a virtual life is far better than physical life or vice versa. Also, there are no major evidences saying the virtual life of every individual is bad. If a person is able to fulfill his or her basic human needs in a virtual world, then there is no issue in that. The ultimate aim of human being is to stay happy.

5.8. CONCLUSION

At the end, it is concluded that understanding the human being role in VE is very important in order to design the proper VR technology and make it better by continuous improvement. There is a need to have a comprehensive understanding about how human react with respect to a particular situation in VR and what are their performance efficiency in VE.

In addition, it is also equally important to have knowledge about the functions of VR for human being. For instance, what are the applications of VR and its role in entertainment and training of people. Having an understanding of all these will help in making these experiences more marvelous.

There is an important relationship between VR and human being, and it should be learned continuously for gradual progress and improvement.

REFERENCES

1. Barbier, L., & Fointiat, V., (2020). To be or not be human-like in the virtual world. *Frontiers in Computer Science*, 2.
2. Kim, M., (2015). *The Good and the Bad of Escaping to Virtual Reality*. [Online] The Atlantic. Available at: https://www.theatlantic.com/health/archive/2015/02/the-good-and-the-bad-of-escaping-to-virtual-reality/385134/ (accessed on 29 April 2021).
3. Lewis, T., & Tech, M., (2021). *Human Factors in the Design of Virtual Reality Instruction*. [eBook] Montana Tech Library Digital Commons @ Montana Tech. Available at: https://digitalcommons.mtech.edu/cgi/viewcontent.cgi?article=1186&context=grad_rsch (accessed on 29 April 2021).
4. Li, Y., Huang, J., Tian, F., Wang, H., & Dai, G., (2019). Gesture interaction in virtual reality. *Virtual Reality and Intelligent Hardware, 1*(1), 84–112.
5. Macwan, U., (2017). *The Impact of Virtual Reality Technology on Human Lives*. [Online] Finextra Research. Available at: https://www.finextra.com/blogposting/14870/the-impact-of-virtual-reality-technology-on-human-lives (accessed on 29 April 2021).
6. Pan, X., & Hamilton, A., (2018). Why and how to use virtual reality to study human social interaction: The challenges of exploring a new research landscape. *British Journal of Psychology, 109*(3), pp.395-417.
7. Sciencedirect.com. (2021). *Virtual Humans: An Overview | ScienceDirect Topics*. [Online] Available at: https://www.sciencedirect.com/topics/computer-science/virtual-humans (accessed on 29 April 2021).
8. Simeone, A., Speicher, M., Molnar, A., Wilde, A., & Daiber, F., (2019). LIVE: The human role in learning in immersive virtual environments. *Symposium on Spatial User Interaction.*
9. Stablekernel.com. (2021). *How Does Virtual Reality Impact the Human Condition? | Stable Kernel*. [Online] Available at: https://stablekernel.com/article/virtual-reality-impact-human-condition/ (accessed on 29 April 2021).
10. Stanney, K., Stanney, R., & Kennedy, R., (2018). *Human Factors Issues in Virtual Environments: A Review of the Literature*. [eBook] Available at: http://web.mit.edu/16.459/www/Stanney.pdf (accessed on 29 April 2021).

11. Thalmann, D., (2003). Virtual humans for virtual reality and augmented reality. *IEEE Virtual Reality, Proceedings*.
12. Thalmann, D., (2021). *The Role of Virtual Humans in Virtual Environment Technology and Interfaces*. [Online] SpringerLink. Available at: https://link.springer.com/chapter/10.1007/978-1-4471-0259-5_3 (accessed on 29 April 2021).
13. Veigar, P. H., (2021). *Virtual Reality Will Bring us Closer to the Human Condition*. [Online] Vox. Available at: https://www.vox.com/2016/8/9/12371460/virtual-reality-human-potential-eve-online-games-theater (accessed on 29 April 2021).
14. Veljović, I., (2021). *Do We Live in Virtual Reality?* [Online] Medium. Available at: https://medium.com/@AmazonkaIV/do-we-live-in-virtual-reality-ebbf37c7f4f0 (accessed on 29 April 2021).
15. Zhang, Q., Zhu, W., & Zhu, Q., (2019). Real-world hand gesture interaction in virtual reality. *Journal of Physics: Conference Series, 1229*, 012027.

CHAPTER 6

Enabling Technologies in Virtual Reality

CONTENTS

Virtual reality (VR) has become part of today's world-leading to its future. Technology in the computerized world has been reaching new heights with this kind of innovation and solutions. The following chapter illustrates about the technology of VR with its modeling features and their effects. It not only briefly explains about its vision and displays but also about its overall tracking systems. Furthermore, it discusses about the effects of its audio system and their history leading to its different affairs as well as feedback.

6.1. INTRODUCTION

Virtual reality (VR) is reliant on the following technologies:

- Real-time 3D computer graphics;
- Wide-angle stereoscopic displays;
- Viewer (head) tracking;
- Hand and gesture tracking;
- Binaural sound;
- Haptic feedback;
- Voice input/output.

The above-mentioned list is used according to the situation, as first three are mandatory but 4th one is used under some circumstances and today in the market 5th and 6th is becoming more important, as it is mostly used by the researchers in the field. The last item which are mentioned in the above list is not considered as the key technology that characterizes VR.

To describe the difference between such system and the conventional computer systems is important. Virtual system is not just bigger, faster but it is considered as the better version of a conventional computer graphics (CG) system. The user is like an external observer in the conventional CG systems, who looks in through a window at some synthetic environment. In the synthetic combination describe above transforms the user into a participant.

There is a difference between the VR experience and the other conventional graphics experience, as the most important contributors to this transformation are the wide field of view (FOV) (greater than 120°), and the binocular and motion parallax achievable with a stereoscopic, head-tracked display and this transformation is a dramatic one.

6.2. 3D MODELING AND COMPUTER GRAPHICS (CG) IN VIRTUAL REALITY (VR)

Areas like entertainment, education, and training, manufacturing, medical, and rehabilitation are the ones in which VR is being widely used, and apart from these areas, there are many other areas in which VR is used. From the year 2018 to 2023, there is an expectation for VR revenue that the compound annual growth rate will grow more than 50%.

In the next 5 years it is expected that education and training will be the leading sectors. Along with the haptic feedback, participants can also interact with virtual objects in VR, as it not only provides immersive stereoscopic visualization of virtual environments (VEs) and sound effects.

It does not matter about the application which is to be applied by VR, to enhance the engagement of participants the visualization and CG are critical and which will result in increasing the education and training effectiveness. But it is not an easy task because the demand and expectation from VR is very high and due to the artist sense of design engineers.

3D modeling and CG techniques drive the explosive growth and becoming crucially important in the recent years because of the era of digital information technologies. For several decades 3D modeling and CG techniques have been developed.

There are many other applications where this technique can get applied, such as artificial intelligence (AI), big data analytics, etc. As, many people think that the techniques are only applied to the development of virtual models for computer simulation, VR, augmented reality (AR), mixed reality (MR), etc. (Figure 6.1).

Figure 6.1: 3D modeling and computer graphics in virtual reality.

Source: Image by Soft8Soft.

As, the development of VR technology has been done several years ago, but with the development of fifth-generation, the 5G network of the mobile has bloomed the 5Vs of the data flowing including volume, variety, and value.

Due to the greater expectations in computer graphical effects, realistic 3D models and infection of VEs, the computation requirements and the flow of big data in VR is very demanding, whereas some people think that this demand is only because of real-time interaction, wireless connection, data inter-exchange.

6.2.1. Modeling and Texture Painting Tools

Into the several key procedures, the development of the VR models is divided. Virtual scenes used in the VR program is created by the VR models. Flow chart of the VR program development is shown in Figure 6.2.

Including modeling, texture painting, and VR programming, the development process is developed into three major steps. To create the object 3D geometries virtual models are firstly modeled using 3D modeling tools.

The next step after the completion of the 3D modeling process is that by using CG techniques including materials painting, texture mapping, etc., the models are being rendered. The software which is required for this process is the 3D modeling software.

Then after this for the development of the VR computer program the 3D models, including the corresponding graphical UV texture maps have to be imported into the game engine. Otherwise, with the help of separate professional software the texture painting and rendering processes can be performed.

Then after this, the texture maps are used as the input of the game engines, which are included in 3D models. Into various file formats depending on the compatibility between the software 3D models and texture files can be exported.

There are some commonly used file formats of the 3D models that are FBX, OBJ, STL, etc. To provide interoperability between digital content creation applications, FBX is used which is a proprietary file format developed by Autodesk. As the part of game wares FBX is commonly used and it is also recommended in the development of the VR program (Figure 6.2).

Figure 6.2: The flow chart of the VR program development. The development process includes modeling, texture mapping, and VR programming.

There some 3D modeling tools such as ZBrush, Blender, SketchUp, AutoCAD, SolidWorks, 3Ds Max, Maya, Rhino3D, CATIA, etc., which exists today.

Modeling tools are professional, and they are basically used for the industrial application. Some advanced features are provided by these tools such as computer-aided engineering (CAE) for performing analysis. Along with this, these tools are used in computer-aided design (CAD), additional manufacturing and 3D printing.

Traditionally, translation of CAD file into VR format directly by a downstream process CAD tools are used. However, complex, and highly detailed CAD data is provided by using the CAD tools. It is common in engineering design and other industries, which makes it translates into excessively large VR models.

In an acceptable level, this makes the models difficult to maintain the sped of computation. In this case, to allow real-time interaction by reducing the complexity of the models which makes the modeling process complicated.

Not only the 3D modeling features provided by the computer-aided industrial design (CAID) tools, but in various industries it is used like 3D orienting, animation, gaming, architecture, and for digital production industrial design is used.

With improved freedom of creativity compared to typical CAD tools provide designers. A format that can be read for AM to speed up the prototyping process is the format in which the 3D model can be saved. This is done with the purpose of providing more time to the designer so that they can focus on the design processes.

For sketching, design, and modeling for designers CAID provides a larger flexibility. The main purpose behind this is to create flexible models to meet the extensive demand of visual realism in VR nowadays.

6.2.2. Modeling and Computer Graphics (CG) in Virtual Reality (VR)

There is an approach which is followed to create 3D models in VR for real-time interaction. The approach is to perform optimization to reduce the complexity by minimizing the mesh size of the models.

Though, there is a drawback attached with this approach that the visual realism of the models may be affected. Therefore, in this some essential modeling and CG techniques which can be applied to create 3D models in VR.

These techniques are used to keep the visual realism effectively without the need of additional modeling procedures, so they are used not only to reduce the mesh size of the models.

In order to make good models there some of the fundamental techniques that should be understood. In the 3D modeling, shading is one of the key techniques. Including flat-shading and smooth-shading there are other several approaches to perform mesh shading.

By polygons and truly curved objects, most of the models represented and are often approximated by polygon meshes, as seen in Figure 6.3. While rendering the models these polygons appear as series of small, flat faces in Figure 6.3(a).

This will result in increasing the number of faces and vertices of the models; thus its complexity therefore is not desired in VR applications. Applying the auto smooth filter to quickly and easily changes the way the shading is consider as the easiest way to generate the visually smooth model. The way of shading by calculating across the surfaces, giving the illusion of a smooth surface is simply changes, as mesh shading does not actually modify model geometry (Figure 6.3).

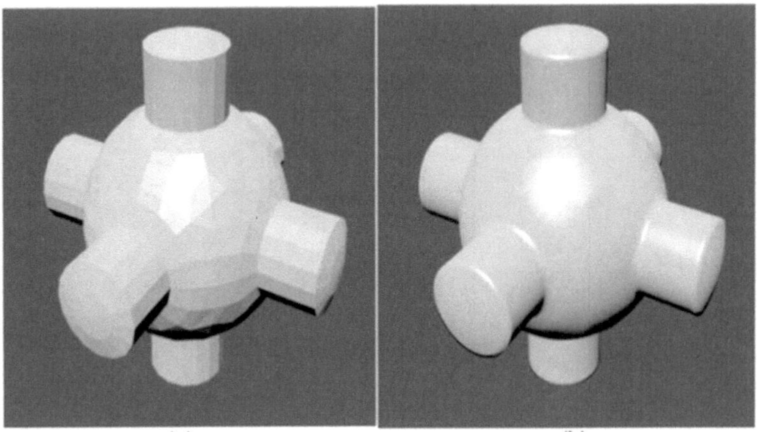

(a) (b)

Figure 6.3: Example of 3D mesh model rendered: (a) flat; and (b) smoothed using subdivision surface.

Source: Image by Intechopen.com.

The mesh non-destructively is created by the shading approaches by calculating the faces normal. On the other hand, in some cases, mesh editing tools such as bevel, subdivision, loop cut, etc., may need to be applied at the edge to create better visual effects.

Figure 6.4 shows with 20 segments, 2 segments and 6 loop cuts (from left to right) the 3D models applied. With the number of bevels segments, the visual effects look similar, but the mesh size increases significantly.

The modeling process is considered more easier, as the bevel modifiers with 2 segments and can create a similar effect with 6 loop cuts. It is preferable that the 3D models to look nice and smooth, although 2 bevels and 6 loop cuts are more effective. Sharper edge will be created by the loop cuts The loop cuts will create a sharper edge, therefore in VR modeling bevel segments are more preferred.

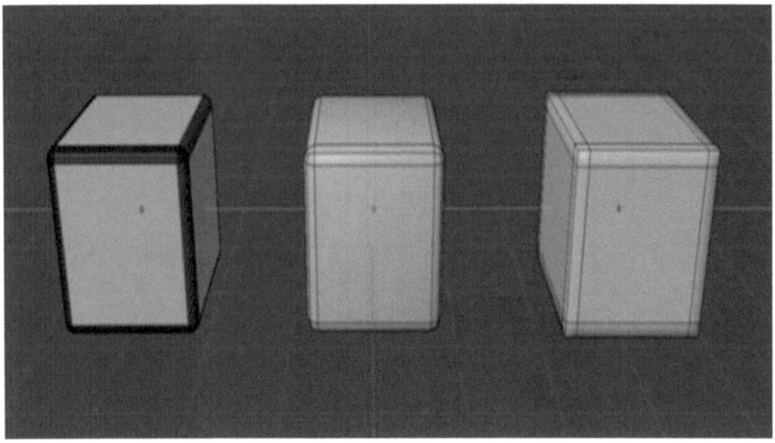

Figure 6.4: Example of using bevel modifiers with 20 segments, 2 segments and 6 loop cuts (from left to right).

Source: Image by Intechopen.com.

6.3. STEREOSCOPIC VISION AND THE STEREOSCOPIC DISPLAYS

An attempt is being made by the human beings to create a display of what their eyes see and their brains interpret since cavemen scratched stick figures on rock walls. Still full gamut of the visual perception system has not been approached by the people.

The assumption has been made to develop a VR display that would present to the viewer an experience that did not fall short of the perceptual detection system that has evolved than what metrics is the design judged by. So, for this, a few rough conservative numbers should be considered.

A horizontal FOV of over 120° without rotating the head is there. In the region of the fovea than in the periphery, the resolution across the FOV is much higher, but to integrate it in high resolution, there is constantly moving eyes around the scene.

For a single eye, some have presented data showing human eye resolution as fine as 20 seconds of arc. However, on average, 10 seconds of arc and at the maximum, 2 seconds of arc stereo acuity has been shown to be (Figure 6.5).

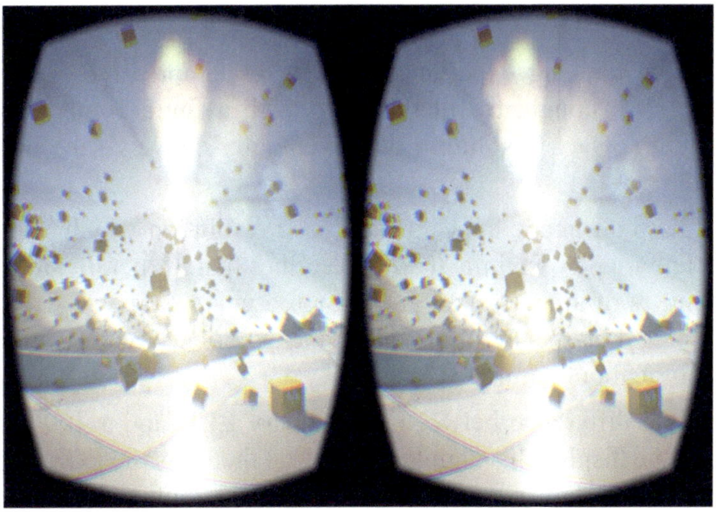

Figure 6.5: Sample screen capture of oculus rift in stereoscopic display.

Source: Image by Wikimedia commons.

A display that would present 120° by 90°, and each eye must be able to see an angular pixel size of 2 seconds of arc then each frame of the display would need about 70 Gigapixels is to be consider.

At the present time to present this resolution economically, there are no electronic display technologies. There are two parameters that describe the perceptual gamut is resolution and FOV. Contrast, cross-talk, color gamut, brightness, distortion, alignment, binocular overlap, accommodation point, and accommodation range are included by others.

Therefore, adopting a set of compromises that lie within the perception gamut is required by designing a stereoscopic display. On the human factors, a very complex problem where the optimizes solution is dependent that vary from one person to another.

And the question which arises is that where does the designer make the trades that will produce the best image experience. For some of these human factors, many people have performed experiments to find an optimum solution space for.

Presenting some of the compromises that can be made is the purpose of this chapter, but not to propose the optimum solution as this optimum solution depends on the application and the subject that is using the display.

6.3.1. Resolution and Field of View (FOV)

An immersive viewing experience that is very impressive is created by the large FOV. Though, to fill this field with a finite number of pixels, the detriments of low resolution will eventually outweigh the benefits of immersion.

How sharp an image appears is angular resolution which is measured in minutes of arc per pixel the metric describes that. Generally, in the range of 1–5 minutes of arc per pixel are in Stereoscopic displays. A very sharp image is created by the arc per pixel, as the experience of viewing displays a resolution of about 2 minutes.

There is a difference between an ordinary LCD and a 400 mm while viewing. In the display this angular resolution then sets the FOV for a given number of pixels. For example, if the display resolution is XGA at 1024 by 768, the FOV is 34° by 25.6° for an angular resolution of 2 minutes per pixel.

Some pupil imaging displays (especially head-mounted displays (HMD)) have compromised the stereoscopic overlap to achieve a larger FOV. For example, each eye may see a 34° FOV, but the field that is common to both eyes may be smaller.

This can give a more immersive experience without sacrificing angular resolution. A large FOV can also increase the fusional range, that is, the amount of disparity that can be fused.

6.3.2. Accommodation-Convergence Mismatch

To give a sense of depth, there are many cues from a scene that visual perception system uses. These include the retinal image size, perspective, shading, and others that can be perceived on a flat screen with one eye.

Accommodation and Convergence are the two very strong depth cues that cannot be presented on a conventional flat display. Looking objects at different depths change the focus in the real world. To objects at different depths, eyes also converge at different angle.

Within a stereoscopic display convergence is what is varied and to stereopsis depth cue is linked. When the viewer is focused and converged at different distances, many have documented the uncomfortable stereoscopic viewing experience that happens.

To find the optimum accommodation distance research has been conducted that allows the largest range of comfortably viewed disparity. By using optics, displays create a virtual image and can also take advantage of this by defining the focus point of the viewer to be more distant and thus can present a wider range of disparity.

Brightness of the display is another metric that is tied into the phenomenon. The pupil contracts and the eye's depth of focus increases, as the display gets brighter, therefore reducing the effect of accommodation-convergence mismatch.

6.3.3. Cross-Talk

It is said that the right eye's image and the left eye's image are completely separated. A unique way of feeding each stereoscopic display has one image to the left eye and its stereo complement to the right eye.

Cross talk is the result of these methods that is a dim image is seen by the left eye intended for the right eye and vice versa. To still maintain fusion, the amount of cross-talk that is acceptable has been shown to be quite large.

User view is also affected by the cross-talk, because how long will the user can view without any discomfort is based on this. Below 0.3% cross talk should be kept, as it has been shown in some researches for ghosting to be imperceptible to the viewer.

6.3.4. Alignment

In creating a comfortable display, the alignment of the stereo pair images is one of the most important considerations. To move horizontally the eyes, have wide freedom but for vertical disparities between images, they have very little tolerance.

Eyes should never have to point beyond parallel is the most important consideration for horizontal alignment. Sometimes, this is known as wall-eyed. There are many factors by which the variation in vertical disparities can be caused.

The obvious cause of vertical disparity is a vertical shift of one image but so too is any distortion of the images. The image position is changed by the distortion as a function of where it is in the FOV. Stereo pair images having common objects with the horizontal disparity, also when the distortion is the same in each eye's image, vertical disparity will result.

Not the entire field view but the portion of the field is view is to be assigned. With respect to rotation, some parts of the field can be fused, and some cannot see when the images are not aligned. Left to right vertical misalignment should be below 10 arc-minutes or the viewer will experience significant discomfort is documented by the human research factors.

There is another misalignment in which the image is presenting in the reverse-stereo, that is, the right image goes to the left eye and the left image goes to the right eye. To novice users, this happens quite often as a stereo image that does not look quite right will be presented by the reverse stereo.

Alignment is said to be critical because even a badly aligned image may be able to be partially fused. A stereo image will be shown by the viewer and a thought will come that it is a good image. Though, eye strain will set in after a short period of time. There will be a noticeable improvement in the stereo quality, and the comfort will be experienced with improvement in image alignment.

6.3.5. Other Metrics and Overall Image Quality

With a terrible stereo image stereo can be perceived. Anaglyph is considered as the one of the most common methods of displaying stereo images. In this method, one eye sees a red image, and the other sees a blue image.

And in this case, the brightness of the image and the color rendition are severely reduced, and the cross talk can be high. With the large amount of vertical disparity, some can fuse the stereo images. The way of people seeing stereo having a large range of differences, but all those stereo images have low cross talk and are bright and uniform are easier for more people to see and can be viewed for a longer period without discomfort that are well aligned.

6.4. HEAD AND EYE TRACKING

With respect to a global zero-point (origin) and has built-in conventions for directions (axes), the synthetic environment that the VR user (called a Cybernet by some) enters is defined. With respect to the global coordinate system, the objects in this environment are defined, and with respect to the same coordinate system, the user has a position and the orientation.

Viewing frustum means the sub-volume of space defined by the user's position, head orientation and FOV and what the user can see in that world depends on this. It is different from the human visual system, but like a movie camera, as in human visual system user can move the eyes independently of head orientation.

Objects which are outside the FOV are still in the environment, and till the time user is not looking in that direction, they are not visible. Translation and rotation of virtual head is required so that VR user can look around in the environment (Figure 6.6).

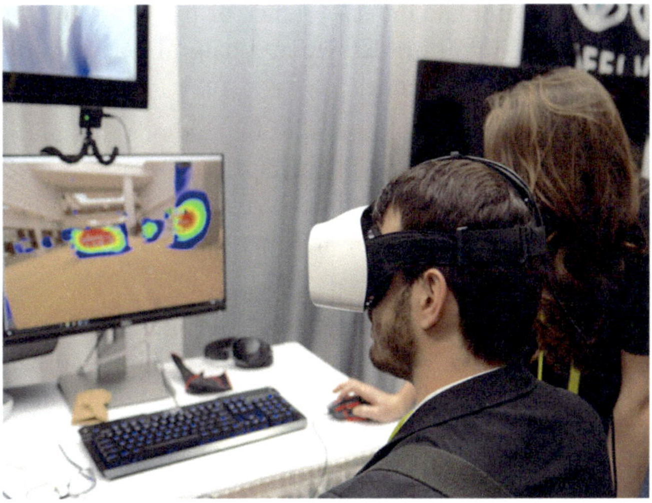

Figure 6.6: VR head-mounted display with eye-tracking.

Source: Image by Flickr.

These so-called viewing transformations were attached to a mouse in conventional CG or in other similar devices. It is said to be functionally satisfactory, as it breaks the coupling between the visual system and kinesthetic senses.

The VR user's kinesthetic senses and visual system can be re-coupled with the head tracking. The display is updated, as the VR user turns his or her head. So, the perceptual fidelity of our system is increased by it.

In addition to the binocular parallax, this the added advantage of the head tracking and from the user will get the motion parallax that is when the head is move from side to side than the side of the objects is seen around.

Depth perception is also improved by this extra visual cue. As, most of this work is not easily applicable to VR at the moment and there has also been considerable work on eye-tracking. With a better description of the viewing frustum and decreases the gap between real visual world vision and our VR model of vision could be provided by the information from eye-tracking.

A user's head position and orientation are monitored by the software application known as head tracking. To help and improve human-computer interaction (HCI), it is often used alongside face and eye-tracking.

In VR and AR to simulate the experience of freely looking around head tracking is often used. In an immersive and natural way to look around in VEs, it allows a user to experience.

For head tracking, there are a number of methods that are being used. Head tracking responsiveness and the screen quality are some of the most significant user experience differentiators between high-end headsets, like Oculus Rift, and low-end headsets and smartphone holding designs like Google Cardboard.

Phone accelerometers and gyroscopes are the ones on which devices that use smartphone often rely on. With precise sensor high end headsets have more accurate tracking and along with this other system including infrared LEDs, cameras, and magnetometers.

Real-life experiences can be simulated in AR or VR because of head tracking. For a more engaging and immersive user experience, it can fool the brain even better than standard viewing. Though, for a prolonged period can cause side effects tracking input lag and looking at screens.

Simulation sickness can be caused with head tracking lag and lower refresh rates on screen. Simulation sickness is a form of motion sickness result in headaches and nausea.

Fields like security, gaming, and medicine in which the head tracking is being used. For computer-aided design, 3-D modeling, and general hands-free computing to improve computer accessibility it can also be used.

6.4.1. Positional Tracking (6DoF)

Oculus Go, Samsung Gear VR and Google Daydream View are the mobile VR headset which only have rotational tracking (3DoF). A user can look up or down, to either side or tilt his head. Leaning of the head's position is not be tracked. It will feel like the entire virtual world will move with you. 3DoF controllers are similar, rotation-only, and they essentially act as laser selection pointers.

It does not allow you to move around the virtual world physically, or to interact with it with your hands directly, and this can be acceptable for seated content (Figure 6.7).

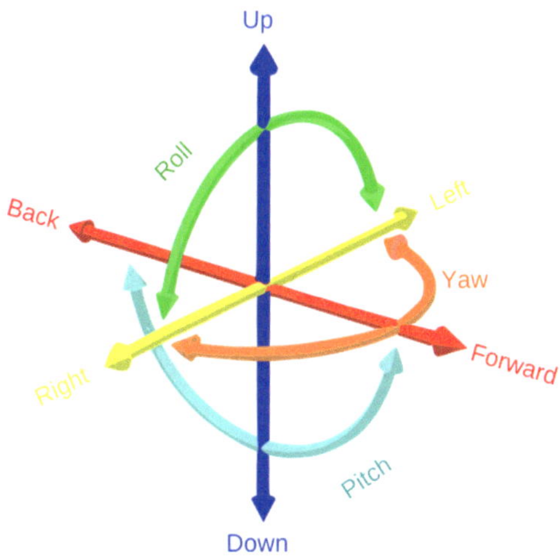

Figure 6.7: An illustration of 6 DoF in virtual reality.

Source: Image by Wikimedia commons.

On PC and console VR, and now in high end standalone VR, the headsets feature positional tracking (6DoF). Allowing the user to move around in the VE is done by positional tracking. A walk around the whole room in VR can be done, if the tracking volume is sufficient.

A user can directly interact with virtual objects with your hands since they can move them through virtual space by moving your hands in the real world, when controllers are also 6DoF.

6.4.2. Tracking Systems

With microscopic electromechanical gyroscopes rotational tracking (3DoF) is always done. But to enable positional tracking (6DoF), different companies use different technologies. Till now, there is not any common industry-standard, but someday a common industry standard will exist. About which techniques are right, companies have different ideas.

Balance cost, ease of setup, tracking volume, controller tracking range, and modularity are the various tracking systems.

6.4.3. The Common Base: Dead Reckoning

Truly correction systems below are described by the optical system. Microscopic electromechanical accelerometer is the primary shared tracking method of all these systems. At 1000 Hz, these accelerometers typically run.

The working of this is explained as the accelerometers do not read position, or even velocity, they read (as the name suggests) acceleration. But over time anyone can take the integral of acceleration and get velocity, as remember from the calculus.

User will get the position if the integral of velocity values over time is taken by the user. To determine change in position by using this is called dead reckoning. According to the moment it depends how every VR headset and controller tracks.

So, after this, there is a question why anything else is needed, and the answer to this is that accelerometers are imperfect providing noisy data. When the smallest error is magnified, it means integrating of data is done twice. In the real world this means that accelerometer-based positional tracking drifts to infinity within a matter of seconds.

By providing a reference the purpose of VR tracking systems is to correct this drift. The purpose always remains the same even when each tracking system does it differently.

6.5. HAND AND GESTURE TRACKING

To interact with the computers now the keyboard and mouse are the most conventional way. For text-based system keyboard is great. For real-time 3D graphics, it is useless. Similarly, while providing an excellent way to interact with a 2D space, the mouse is used.

There have been a few conventional devices that have provided 6° of freedom (DOF), notably the Space digitizer and the Spaceballs.

To provide a means of measuring points on a physical model and enter them into a CAD database the Space digitizer was designed. The position and orientation of the pen with respect to the control unit is measured with a sensor attached, and it is a like stylus (Figure 6.8).

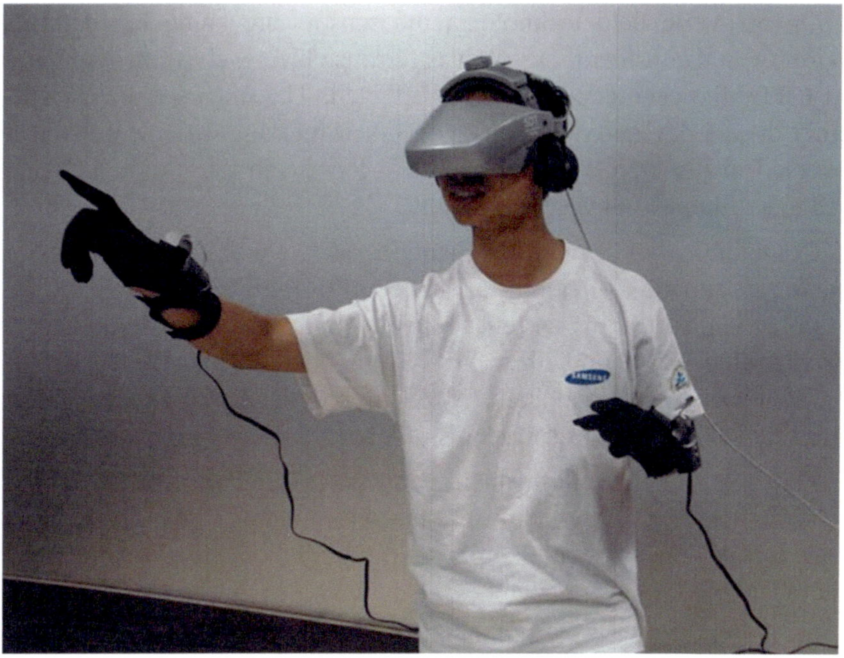

Figure 6.8: Hand and gesture tracking in VR.

Source: Image by Flickr.

To calculate the position and orientation the sensor, the Polhemus, uses perturbations in orthogonal magnetic fields (caused by an AC solenoid in the pen). To track a position and orientation in 3D space, the required information is provided in this theory, but there are several problems attached with the Polhemus sensor.

Being a hemisphere of approximately 30-inch radius from the transmitter, the working volume of the sensor is very small. From 0.13 inches, its static accuracy in that hemisphere ranges and if the sensor is less than 15 inches from the transmitter, to 0.25 inches if the sensor is up to 30 inches away. About 0.85° the orientation accuracy is.

Ranging from 0.09 to 0.18 inches the usable resolution of the device is also poor. At 60 Hz each sensor can return positional information. Though, in the same environment, several sensors are being used, and they must be time-sliced to avoid the magnetic fields interfering. There is a reduction in the maximum update from 60 Hz to a barely acceptable 20 Hz for a typical setup of three sensor.

On magnetic field technology, the sensors are being based, and by environmental considerations such as ferrous objects, electronic equipment, and CRTs, the accuracy is strongly affected. The magnetic fields will be highly distorted, resulting in very non-linear measurements which must be corrected in the typical environment. The noise in the returned data will need to be filtered and very significant.

Combination of latency (acquisition rate in the analog domain) and update rate (number of samples per second of digitized data) is known as phase lag. Up to 100 ms current Polhemus trackers have a total phase lag. For real-time interaction, this causes serious problems. By the graphics latency this is exacerbated that is added after all the tracker phase lag. Using good predictive filters, the problems of latency can be partly compensated.

For VR today the Polhemus is the most popular sensor, despite of shortcomings. Under the palm of your hand, mounted on a low stand, the Space Ball is a rigid sphere that fits comfortably. Force that is applied to the sphere is measure by strain gauges.

To measure translation in three dimensions and three angles of rotation, there are enough sensors. To provide the necessary functionality for interacting with objects in 3D, this is clearly sufficient. The main disadvantage of the space ball is it does not move.

Instrumented glove is the most popular VR input device. The classic example of this is the VPL data glove. It has a Polhemus sensor attached to the back of the hand and ensures sensors for each finger and the thumb.

To find the position and orientation of the hand (through the Polhemus) at any instant (the system queries the glove controller for its current data) is allowed by this system. Joint angles of each finger and the thumb is measure by the flexure sensor.

This data can be used for gesture-based input to the system when combined with recognition software. This is like to enter commands into the computer by using a form of sign language.

To recognize gross motion gestures of the hand (say waving goodbye) also to be used for interaction is a good opportunity. To recognize gross motions of a mouse and should be readily extensible to a Polhemus the GRANDMA system developed at Carnegie Mellon University has been used successfully.

Data Glove, including a high noise level, slow sample rate, noticeable lag is facing some of the difficulties, and at regular intervals, the flexure sensors need to be re-calibrate. By the new competitors to the Data Glove such as the Cyberglove from Virtual Technologies, some of these difficulties have been successfully addressed.

Superior flexure measurement system that provides greater resolution and better repeatability the Cyberglove has. For more information about the user's hand is provided by the addition of extra sensors.

Intolerance of Polhemus to metallic objects in its operating envelope has been addressed by a new sensor. The Bird, from Ascension Technologies, while problems with noise, slow sample rate and lag may well be addressed by a new model from Polhemus called FASTRACK. It claims a phase lag of only 4 ms and an update rate of up to 300 Hz.

Exos has developed an exoskeleton (Dexterous Hand Master™ that provides high precision joint/flexure measurement. Exos has also Inco orated a tactile feedback device (Touch master™) based on a low-cost voice-coil oscillator.

6.6. COMPARING TOUCH-BASED AND HEAD-TRACKING NAVIGATION TECHNIQUES IN A VIRTUAL REALITY (VR)

In the recent years, in surgical simulation-based training, VR technologies started gaining momentum by allowing clinicians to practice their skills before performing actual procedures. Primary operative task which is to be taught required the complete focus of the simulators.

Task need to perform by the user, so little attention should be paid to the secondary task. The perfect example of this is that changing his/her point of view while manipulating the surgical instruments. Mainly for those tasks, it is not clear that how to design appropriate interaction techniques and how on such systems the fidelity of these interactions can impact the user's performance.

Two viewpoint changing techniques is compared in this chapter and these two techniques are having two different levels of interaction fidelity during needle insertion in a semi-immersive VR (SIVR) biopsy trainer. Based on observing clinicians performing actual biopsy procedures, the techniques were designed.

Tracking the user's head position (high interaction fidelity) the one on which the first technique is based and on the other hand second technique is known as touch-based in which the user is utilizing his/her non-dominant hand fingers to manipulate the point of view on a touch screen (moderate interaction fidelity).

To investigate the impact of the interaction fidelity of the viewpoint changing task (secondary task) on the user's performance during the needle insertion task (main task), a user study was carried out. There was a trial for a needle insertion task and the 21 novice participants were asked to perform several trials for it by using the navigation technique.

To compare the task performance and user experience for both techniques, objective and subjective measures were recorded. As per the result, the user task completion performance is improved by the touch-based viewpoint changing technique, during needle insertion while maintaining a similar level of needle manipulation accuracy as compared to the to head-tracking technique.

So, the result suggests that while designing surgical trainers, it is not always necessary to have the high interaction fidelity. The importance of designing appropriate interactions for secondary tasks is also highlighted by this because in this, the user's primary task performance in VR simulators can also be influenced.

There is a minimally invasive surgery (MIS) procedures which is to be followed by the surgeons to view the operating field on a monitor while manipulating surgical instruments inside the body. For patients, these techniques have many advantages like less pain, smaller scars, and faster recovery.

Many challenges have been presented for clinicians, such as must having the ability to master new complicated skills like translating the two-dimensional (2D) screen images to a 3D working area. Legal, ethical, and safety issues are there with training on live patients, so the traditional apprenticeship model based on training on live patients is not only insufficient.

For a new learning model simulator emerge as medium and come in the way to solve those issue. Certainly, for both trainee and patients, the risk is being reduced by allowing the training and the assessment of skills in a simulated environment prior to the real-world exposure.

6.7. BINAURAL AUDIO

For headphone listeners, binaural sound technology allows the creation of immersive spatial audio experiences. It will help in enhancing the programs which are listed on the headphones and create immersive interactive experiences including in virtual and AR.

In the way user hear the world binaural or binaural 3D audio is audio captured identically in the same way. The exact location of every sound is captured when audio is captured with a binaural microphone like the Hooke Verse and along with this source of sound and where it is in relation to the recordist upon capture (Figure 6.9).

Figure 6.9: Binaural audio in VR.

Source: Image by Unsplash.

Two omnidirectional microphones is consisted by a binaural microphone wherein one mic is positioned very precisely within each ear. A recordist is capable of recording audio identically to the way they hear it When a microphone is placed within the pinnae of each ear.

6.7.1. How Does Binaural Audio Work?

There is nothing to get confused in the binaural sound with conventional stereo sound. Allowing for localization to the left and right upon playback is done by stereo audio. And in the case of binaural, sound sources can be localized by a listener in front of them, behind them, above, and below. In the natural ear spacing or head shadow of the head and ears stereo does not factor.

These things happen naturally as a person listens, generating their own ITDs (interaural time differences) and ILDs (interaural level differences). Stereo recording is not capable of capturing these ITDs and ILDs.

Between the bodies and the environment around binaural techniques simulate the hearing cues created by acoustic interaction. To introduce these cues and give the impression that a sound source is located outside of the head at a given location in space, audio signals are being filtered.

> Binaural beats in the alpha frequencies (8 to 13 Hz) are thought to encourage relaxation, promote positivity, and decrease anxiety. Binaural beats in the lower beta frequencies (14 to 30 Hz) have been linked to increased concentration and alertness, problem solving, and improved memory.

To the inaccurate cues system appears to be sensitive, and this is common in binaural filters to create an unconvincing spatial impression as well as poor sound quality. Cues are different of every individual as it is created according to the unique body shape, so this means every person is having an individual pattern of hearing and these cues change with the movement of the listener.

For convincing effect natural binaural reverberation is also important. Specialists of the equipment and careful measurement is required for achieving high quality binaural sound. New techniques based on a better understanding of human hearing is required to adapt this for broadcasting.

6.7.2. The History of Binaural Audio

In the late 19th century, the binaural audio was invented, and at that time amplifiers did not exist. At that time only way to listen to electrical audio signals was through a telephone receiver earpiece, and every radio came with a pair of headphones until the mid-1920s.

But afterwards, with the introduction of the moving coil, everything has been changed. And by the end of the 30s, paving the way for stereophonic sound was being demo-ed for movie soundtracks. Till today binaural slips to the background.

In the today's world every smartphone comes with the earbuds and everyone is having craving for immersive experiences. And creators in every sector (VR, music, podcasts, gaming) are tapping the limitless potential of binaural.

The first binaural audio system was invented by Clement Adder in 1881, and he is the person who has established the telephone network in Paris in 1880. Adder also installed an array of paired microphones at the edge of the stage of the Opera Garnier and transmits the audio via two phone lines, one for each ear to listeners located two miles away. Afterwards binaural system is installed in the theaters through Europe. The cost of listening at that time was 50 French centimes per 5 minutes.

6.7.3. Why It Matters?

In the past few years, large growth has been seen in the number of people listening to programs over headphones. This is done with the arrival or invention of the smartphone, fast data networks and streaming services such as the BBC iPlayer.

In stereo, most audio is heard over the headphones, and that will be the same content which is being played over the loudspeaker. A flat impression with sounds coming from inside the head can be given by listening to stereo programs over headphones.

To create a richer sense of space in program sound and giving a more exciting and immersive listening experience binaural techniques can be used. For loudspeakers working is going on for the 3D sound, but that kind of experience to listeners' headphones can get by binaural technology.

To new interactive content experiences including virtual and AR, this can be applied and along with this to traditional program content.

6.8. BINAURAL AUDIO FOR VIRTUAL AND AUGMENTED REALITY (AR)

Through headphones, immersive VR and AR experiences rely on 3D sound that is as close to real-life as possible. In collaboration with the google, this project looks at research undertaken at the audio lab, which has set the global benchmark in binaural surround sound for VR (Figure 6.10).

Figure 6.10: Binaural audio for virtual and augmented reality.

Source: Image by Unsplash.

6.8.1. The Issue

Renewed interest is experienced by the creation of immersive 3D sound over headphones. To technical advances in ultra-high-definition video displays and interactive VR headsets this can partly be attributed, but a surge in production support for consumer-based 360° visual and audio content generation and consumption is also being seen.

With the use of headphones, binaural surround sound is delivered and it is commonly used to accompany such great immersive display. Forming realistic, or hyper-real, sound fields that are experienced with good externalization, localization, and sound quality is a challenge for it and for mass audiences with uncompromised sound quality this research project looks at enabling binaural reproduction.

6.8.2. The Research

There some special filters by the name head-related transfer functions (HRTFs) and their role are that 3D audio can be created by filtering recorded sounds. Interaction between the head, torso, and ears of a listener and a sound source at a given angle relative to the head is described by these. With the use of probe microphones HRTFs measures in a subject's ears or from binaural dummy head microphones.

The basic purpose of the approach is to treat the measured HRTFs as virtual loudspeakers equally spaced around the head. To create the signals that feed the virtual loudspeakers to create the immersive soundscape, a 3D audio rendering technique is used known as Ambisonics.

There is an advantage with this technique that the entire sound scene can be translated to counter head movements. The sound will remain stable in 3D space as it does in real life if a listener moves their head in a headphone-based and motion-tracked VR experience.

As part of the Engineering and Physical Sciences Research Council (EPSRC) funded Spatial Audio for Domestic Interactive Entertainment Project (SADIE), this work was started. The main purpose for which the project has been started is that to improve the immersive sound rending in the home. Binaural filters have been there to support and their data sets are also being measured for it in creating new algorithms to improve VR audio rendering.

YouTube 360 is the streaming service and the question which comes to every mind is how 3D audio sound scenes can be best compressed for streaming services. Optimal bit rates and encoding of conditions for binaural-based ambisonics have been evaluated by York's audio lab to be delivered through Google.

6.8.3. The Outcome

Into Google's VR pipeline SADIE binaural filters have been integrated and at all stage's VR pipeline influencing the audio rendering of VR content creation. Google Resonance Audio is the heart for them.

Google has shipped more than 10 million Cardboard VR headsets to understand the impact to date. Over 1,000 applications were developed with 25 million installations made worldwide over the first 19 months of the Cardboard platform. In May 2016, with the Daydream VR platform, whose VR mode included apps such as YouTube 360 Google followed this up.

To improve spatial audio quality through the browser, specifically looking at the optimal bit-rates and compression for streamed spatial audio delivery whilst preserving spatial fidelity and timbre the audio lab continues to work with Google.

6.9. SOUND IN VR/AR

VR and AR applications will be the next generation of medicine and healthcare. But now let us give a thought to the key aspect of any VR/AR environment sound.

VE is considered as unrealistic without sound. The users will feel fully immersed in the VE only when the sound is getting right in the VR. The spatial sounds need to match the spatial characteristics of the visuals to ensure full immersion in VR systems.

For example, if a car is moving away in the VR environment than the expectation will be to hear the car is moving away. If this not happen than it means the whole impression of being in VR is immediately lost.

With use of loudspeakers which are being placed all around the listener's head is the only way to recreate an immersive sound field. As previously, it has been seen that how a large array of 50 loudspeakers is used to reproduce complex sound scenes with sound sources located all around the listener. By extending the techniques used in stereo recording and playback, recreating a sound scene over multiple loudspeakers is achieved.

6.9.1. Stereo Sound

If there are two loudspeakers than the perceived position of a sound source can be moved to anywhere along the horizontal plane between the two loudspeakers. By increasing the amplitude level of the left loudspeaker and lowering the amplitude of the right loudspeaker, the sound to the left side can be panned.

Through both loudspeakers, if the sound is played at the same amplitude level, then it will be heard as if coming from directly in between the two.

This technique of amplitude panning is used across an array of multiple loudspeakers to reproduce three-dimensional (3D) surround sound, and this is basically based on moving sound sources between loudspeakers to scale up.

A large number of loudspeakers arranged in a sphere can produce enveloping sound around required access, but not many people are having those access. The design of the VR/AR systems is basically for personal or home use and relies on headphones to deliver surround sound to the user.

6.9.2. Spatial Sound

To listen to all the sounds around, everyone has to listen very carefully by closing their eyes for a moment. So, after listening carefully, some questions come to every mind that is What can you hear? Where are those sounds coming from? What can you hear in front of you? What can you hear behind you? What about above you, below you, or to each side? Try rotating your head-do the sounds change?

Hearing system of the user is very well evolved even when the eyes are being closed. And it is evolved to allow the user to locate sounds in the space around. This known as sound localization.

Binaural cues are the one on which rely can be done to localize sounds in space. About the level, timing, and overall tone of the sound arriving at our left ear brain takes the cues information. This is done to compare it with the sound arriving at the right ear. In each ear differences between the sound helps to work out where the sound is placed relative to our own position.

The discovery has been done that how sounds are filtered by the outer ear and how sound changes in the ear, mainly the pinna. If the sound place is at right then slightly move the head above, then the wave that reaches the ears from this sound has traveled directly to tight ear but to reach the left side of the ear, it has to travel around the head. The sound is actually get filtered by the shape of the head. This means these are spectral cues in which some frequencies are dampened and the overall tone is altered.

There are two more types of binaural cue which a brain can use:

- ***The Interaural Level Difference (ILD):*** The sound has traveled further to get to the left ear, so it's quieter because it's lost more energy on the way.
- ***The Interaural Time Difference (ITD):*** The sound reaches the right ear a fraction of a millisecond before it reaches the left ear.

6.10. HAPTIC FEEDBACK

The main problem with the VR system is that there is no physical feedback. This is because in this user will not feel the sensation of touching anything (Figure 6.11).

Figure 6.11: A glove powered by soft robotics to interact with virtual reality.

Source: Image by Jacobs school of engineering.

Therefore, there are two issues to be addressed, which are discussed in subsections.

6.10.1. Tactile Feedback

A mean to provide the feedback with the touch or through a skin. For example, array of vibrating nodules under the surface of the glove. This is not an accurate simulation of touch, but it can be considered as it would at least provide some indication of surface contact. The other example for this will be inflatable bubbles in the glove, materials that can change from liquid to solid state under electric charge and memory metals.

Tactile feedback device has been incorporated by the Exos into their Dextrous Hand master. On a low-cost voice-coil oscillator it is based on. By using 30 inflatable air pockets in the glove tele tact glove provides low resolution tactile feedback. Up to 12 psi for most pads and up to 30 psi for a large palm pad, these pockets can be individually inflated.

6.10.2. Force Feedback

To enforce physical constraints, means are provided, and along with this it also simulating the forces that can occur in teleoperation environments. Some devices have been built to provide force feedback:

- MIT force feedback joystick;
- Ceiling-mounted force feedback arm at the University of North Carolina (UNC);
- UNC mountain bike;
- The steering wheel in Atari's Hard Drivin' and Hard Racin' arcade games.

6.11. VOICE INPUT/OUTPUT

It has been seen that even at a personal level today, speech synthesis is common. In a VR environment, there are some facilities which are having clear utility like spoken word can be there instead of text in case of online help and command feedback can be a computer-generated text. In speech synthesis, there is still a requirement for the improvement, especially in the quality of speech.

In the field of voice recognition system, a considerable amount of work has also been done to the point where commercial systems are available off-the-shelf. Though, the budget of current VR is less so the goal of person and accent independent continuous voice recognition is beyond. This mean for there will be a training process to go through for each user. A noticeable gap between each word is required, and it should be taken care by the user carefully.

6.12. CONCLUSION

All-inclusive this chapter leads around the speech and sound systems included in VR and AR. Briefly leading to its overall progress with its advantages and disadvantages, this section provides us with a deep sense of feedback included with its usage.

Moreover, it gives user explanation about the binaural audio and issues which could be harsh in its results and also with better scope of progress. Overall, one could say that VR has become an integral part of our world today, a part of its negative feedback and more scope for improvement.

REFERENCES

1. BBC R&D, (2020). *Binaural Sound.* [Online] Available at: https://www.bbc.co.uk/rd/projects/binaural-broadcasting (accessed on 29 April 2021).
2. David, H. U., (2019). *How Virtual Reality Positional Tracking Works?*. [Online] VentureBeat. Available at: https://venturebeat.com/2019/05/05/how-virtual-reality-positional-tracking-works/ (accessed on 29 April 2021).
3. FutureLearn, (n.d.). *Sound in VR/AR.* [Online] Available at: https://www.futurelearn.com/info/courses/creating-the-amazing-engineering-the-future/0/steps/65304 (accessed on 29 April 2021).
4. Hooke Audio, (2020). *What is Binaural Audio-Hooke Audio*. [Online] Available at: https://hookeaudio.com/what-is-binaural-audio/ (accessed on 29 April 2021).
5. https://www.researchgate.net/publication/341997368_Comparing_touch-based_and_head-tracking_navigation_techniques_in_a_virtual_reality_biopsy_simulator. (2021). *Comparing Touch-Based and Head-Tracking Navigation Techniques in a Virtual Reality Biopsy Simulator*. [Online] Available at: https://www.researchgate.net/publication/341997368_Comparing_touch-based_and_head-tracking_navigation_techniques_in_a_virtual_reality_biopsy_simulator (accessed on 29 April 2021).
6. Mod.studio, (2016). *VR in a Nutshell VR in a Nutshell.* [Online] Available at: https://mod.studio/vr-in-a-nutshell/ (accessed on 29 April 2021).
7. Ricca, A., Chellali, A., & Otmane, S., (2020). Comparing touch-based and head-tracking navigation techniques in a virtual reality biopsy simulator. *Virtual Reality,* [Online] *25*(1), 191–208. Available at: https://link.springer.com/article/10.1007/s10055-020-00445-7 (accessed on 29 April 2021).
8. WhatIs.com. (n.d.). *What is Head Tracking? Definition from WhatIs.com*. [Online] Available at: https://whatis.techtarget.com/definition/head-tracking (accessed on 29 April 2021).
9. York, U., (n.d.). *Binaural Audio for Virtual and Augmented Reality.* [Online] University of York. Available at: https://www.york.ac.uk/research/impact/binaural-sound/ (accessed on 29 April 2021).

CHAPTER 7

Benefits and Applications of Virtual Reality

CONTENTS

The following chapter illustrates about the importance of virtual reality (VR) in every industry like education, health, tourism, retail, as well as business. It discusses different aspects of VR for present as well as future times. It illustrates various pros and cons of VR how it effects psychologically, socially, economically, and globally. Further, it highlights the efficient usage of VR in increasing digital marketing for the future. It also tells that with the invention of virtual technology, things have become much convenient in every industry.

7.1. INTRODUCTION

In the coming future, VR is expected to be an important technology. It is expected that by the year 2020 VR industry will become a staggering $150 billion market. There are many top tech companies that are already doing investment in the VR industry, and almost 80% of the top tech companies have invested in VR and it is also expected that in the future, this number will only continue to grow.

Earlier, this technology is mainly used by the gaming industry, as the gaming industry is considered as the most popular domain, which uses VR, but now the time has changed, and VR technology is no longer limited to this industry alone. Nowadays, it has been seen that the use of VR is massively increased by the various other industries, such as real estate, business, marketing, etc.

Earlier it has been seen that there were many businesses that were not convinced in the past considering the high cost of the VR devices. But nowadays, there are many varieties of VR headsets which are available in the market at affordable rates which means that in the coming future businesses will soon enter the virtual world easily and the virtual will be beneficial for all the businesses.

So, it is said that there will be numerous opportunities for businesses to incorporate them into their daily operations and get the benefits with the increase in the number of people using VR.

7.2. VIRTUAL REALITY (VR) IN EDUCATION: BENEFITS, TOOLS, AND RESOURCES

Old age movies such as "Fantastic Voyage" being released in the year 1966 works around imaginative drama by showcasing the stories like these were shrinking to human size cell is a common phenomenon whereas in the real world it's completely impossible to achieve under the human tendency (Figure 7.1).

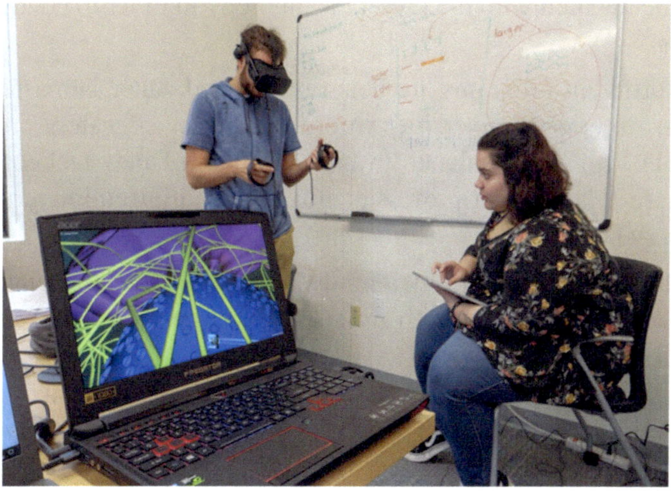

Figure 7.1: Application of VR in education.

Source: Image by MIT Scheller Teacher Education Program.

Moreover, motion pictures showing the human capability to remove the brain clot by riding through the bloodstream increase human aspiration to achieve the same.

Furthermore, with advanced technology such as virtual reality (VR) as well as augmented reality (AR), students are achieving the same imaginative worlds today. With the range to choose from whether it's an outer space or to reach any ancient times they can reach anywhere they want as well as the pathways inside human systems.

There are both pros and cons regarding this new technology. There are different perspectives regarding the same. Some instructors highly support VR and AR technology, whereas others don't due to the high costs being involved. On the one hand, where some teachers consider it as an effective learning device for the classroom; on the other hand, some only see it as a device for enjoyment.

Furthermore, there have been many connectivity issues being reported in Ed Tech report, such as accessible content as well as quality and size of the equipment, which are difficult to maintain.

Regardless of these issues, it is still expected to see the increase in demand for AR and VR technology for future. Therefore, it is expected out of present and future educators to develop their knowledge of virtual systems as well as be updated with the present technology in order to conduct future classrooms.

Many different programs are being introduced, such as master of arts, teaching in American universities in order to make educators progressive thinkers who can lead students towards the path of new technology. The graduates are asked to get professionalized in multidisciplinary approach, which is very beneficial.

Programs like these helps their graduates to get skilled in technical work with real-world experience with the use of this new virtual program and advanced technology.

The program prepares graduates with real-world technical skills using advanced virtual platform technologies.

7.2.1. What Are the Benefits of VR in Education?

In order to define VR, one needs to see how it is separated from AR. On the one hand, AR comes as an educational device to provide appropriate content for the lesson and for projects for students in order to make it more connected and significant for them whereas, on the other hand, VR comes as a 360° device which creates a virtual environment (VE) which feels real and students can have this 3D experience as if they are present there.

Moreover, this VR not only inspire students to discover new academic heights but also gives them immense learning experience with vast imaginations to explore and increase their thinking power.

In order to understand hard academic concepts, AR and VR comes as very helpful tool. This can be understood with an example mathematical student can understand geometry through AR and can see 3D geometric shapes and angles from inside through VR and they can cover it from all the angles.

Apart from academics in education, there are many other advantages of VR, such as knowing about a person's culture and values in the current globalized world. For example, a student can learn about cultures through VR field trips it does not matter where they are sitting, they can connect to the whole world, whether it be Peru or China.

Furthermore, any particular student can improve his or her result with the combination of both VR and AR in education. For example, as cited in the EdTech report published in 2019, students in a mixed reality (MR) biology classroom received higher scores than other students.

Similarly, this advanced technology of AR and VR helps in memory retention and recall. It is shown in the EdTech report how student who learned under the environment of VR resulted in increase in retention in almost of 9%.

7.2.2. VR in Education: Resources and Tips

There are other available resources to make AR and VR tools cost-effective. Any educational system can go for the less expensive tools such as google cardboard which can be easily connected to smartphone and are much lesser expensive and can get easily connected.

Resources like these provide educators with affordable apps and certain free applications such as visiting 360 cities with different locations like Rome and Tokyo. Furthermore, apps like time looper can help students to go through historical platforms such as London world war II and also help educators to plan their lesson with this technology through VR and AR.

There are few tips for teachers to maximize the benefits of VR apart from other resources:

- ***Ensure Ample Physical Space:*** Safety is the most important tool for students, in order to use VR equipments. Often, users ignore the physical space in order to use VR and mostly spin around or blindly move around. To obtain the benefit of VR in education, students need to make sure of their physical environment in order to avoid accident or injury.

Hence, teachers need to be sure about their classroom physical space and should make students at least an arm's length away from each other. Moreover, the teachers can also use the VR content, which can be easily accessed by students just by sitting on their desk.

- ***Supervise and Moderate VR Use in Classrooms:*** The VR needs to be used with closed supervision for school in order to avoid the psychological impact of VR on students. As per the research report with CNN.com, the article provides the result of overused VR by the children and how they have false memories of physical spaces they never visited. With longer tutorial session and less VR education session, this issue can be addressed.

- ***Know When to Use VR in the Classroom:*** It is said that VR cannot replace the human interaction and VR also offers some new opportunities by bringing academic subjects to life. VR is considered as an additional learning tool, and it is fundamentally a social experience.

The use of VR in the classroom by the teachers is totally depends on the subject. There are some subjects like teaching grammar in classrooms by using VR does not make much sense because grammar is a relatively abstract topic.

Apart from this, the topics that are visual and tactile VR may work well. For example, student learning about the historical event or to monument.

There are many cases which can be studied, like there is a physical structure in Greece by the name Parthenon, so to see the architectural detail of its students can virtually walk inside with the help of VR equipment and software.

There many other topics which can lend themselves well to VR. Topics like science, technology, engineering, and math topics also lend themselves well to VR.

- ***Develop a Plan for VR Learning:*** In the classroom, there are many notable benefits of the VR, which spark interest and attention in the students. Educators should develop a structured plan to maximize the use of VR within lesson plans, and along with this, they should guide the students. Development of a structured plan is necessary because students can turn off their own devices.

To ensure best learning experiences, teachers have the responsibility to understand the goal and expectations of the students, and along with this, to achieve the goal, teachers should set the guidelines.

- ***Teach Empathy and Cultural Competence:*** VR perform the magic into the classroom by bringing different places together throughout the world. The benefit of this new view point is that it results in the development of empathy and cultural competence because they take students outside of their normal daily experience. People around the world is different and unique, so with the help of VR and AR students understand people's unique situations across the world.

For example, with VR application, teachers can provide the exposures to the students for different language teaching, as there are people of different cultures with different languages. Along with this, technology advancement and with the use of that technology culturally responsive environments is build, and it helps students to respect cultures different from their own.

7.3. BENEFITS OF APPLYING VR TO BUSINESS

It has been said that there are two main reasons for the leading brands to create VR marketing experiences is engagement and inspiration. These reasons will help the company to be in front all the time in term of technology, and it will also help in drawing audience attention.

7.3.1. Revolutionize the 'Try Before You Buy' Concept

There is a concept named try before you buy which is expanding very rapidly, as more and more retail companies are integrating VR/AR technologies into the customer experience. Basically, it is more beneficial for the consumer, as they can have a look before purchasing it, and it will be considered as a unique consumer experience.

Nowadays, if anyone wants to buy a Volvo car, then the Volvo company will provide a VR test drive on the phone. Lacoste is a brand that has created an AR mobile app that customers could use to try on shoes virtually. That is considered as the perfect marketing combo.

7.3.2. Introduce Established Products to New Audiences

The experience of a quality VR/AR content brings the product presentation to a new level. This way of presenting the products and services is considered as a cool and unique way, and it will help to create a much more interactive shopping experience. It also creates an emotional connection with customers, as it allows customers to consume the content in a more immersive environment.

7.3.3. Prototyping and Design

Earlier developing some products was very costly like 20 years ago if product had to be launched in market, then the launching of product is very expensive, time-consuming, and risky. With the advancement of technology, the process of product development is shifted to the fast process. Now a day's product development is faster and cheaper because of technology.

Time is very important in every industry, so prototyping is considered as the best means to prevent wasting your time. Prototyping provides some special tools and approaches for exploring ideas, furthermore, testing it before excessive resource usage.

It is basically an experimental model which helps to test the idea and avoid spending time and money without a 100% guarantee of desired results. During research, it helps in exploring problem in the process.

7.4. INDUSTRIES THAT WILL BENEFIT FROM APPLYING VIRTUAL REALITY (VR)

7.4.1. Healthcare and Well-Being

It is said that VR/AR plays a very important role in the healthcare industry as it offers doctors some new possibilities for finding innovative treatments. As it has been seen that there are very huge benefits in the VR of medicine.

To give surgeons 'X-ray vision' during complex spine surgery medical operation an augmented-reality surgical navigation system was designed. Earlier anxiety or phobia is not treated well enough but with the VR in medical field could help properly treat phobia or anxiety. Real-life scenarios can be overcome that cause fear.

Virtual walking should be done under the guidance of a coach, and it should be done to overcome a fear of heights. VR pediatrics can also help the doctors to do the treatment of kids without any pain. Also, in rehabilitation and physiotherapy, these digital tools play an important role.

It is said that for immersive education of medical students, there are two great platforms which are known as virtually rendered operating rooms and virtual modeling of organs and human anatomy. When it comes to VR in healthcare, the minimization of risks and more efficient productivity are the two primary goals.

7.4.2. Entertainment and Gaming

The two industry which are known as the inventor of VR and AR or the industry where these two have been applied in entertainment and gaming. These such platforms make unique gaming experiences possible (Figure 7.2).

Figure 7.2: Application of AR in gaming and entertainment industry.

Source: Image by Unsplash.

The great feature is that people can also interact here with each other in a 3D VR entertainment environment during a game. There some chat rooms where the people can hang out, and it has also become an integral part of the game.

Virtual galleries, theaters, cinemas, theme parks, museums included in VR entertainment. VR tools make possible to touch untouched museum. With the use of gloves, anyone could touch which represent different materials, moreover it allows receiving feedback. For blind people, it is a brand-new experience. Furthermore, it allows to live a life like a superstar.

7.4.3. Tourism

In the coming future, it will not be difficult to predict that visualization of upcoming trips will be in high demand, as the popularity of services like Google Street view today. There is various direction in tourism industry where VR can be used (Figure 7.3).

Figure 7.3: Application of VR in the tourism industry.

Source: Image by Flickr.

The first impression is the last impression and traveling is all about impression. So, the application or the website of the tourism should allow the user to make their own virtual tour to their favorite destination.

7.4.4. Retail

It is said that in the retail industry, virtual and augmented technologies are serious game-changers, as they are both online and in-store. The experience is too good that it simulates physical retail stores in online retailing. VR in retail space can have a measurable impact, as a shop design becomes much more intelligent.

Usually, the store is having many sections and many floors, so navigation within the stores is a new way which is offered by AR navigation. Along with this AR navigation help the customer to find the places quickly as the mobile app provides AR routing.

There is another app in the mobile which is very popular by the name AR-overlaid product information. The main feature of this application is that just point phone camera at products and get the detailed information.

7.4.5. Education and Training Programs

In education and training, virtual training simulates the real environment, which is considered as the one of most crucial benefit. For example, if the history classes will be being taught with animated timelines function, visualized chronicles events, or on the other hand some attractive geography lessons with attractive trips across the continents.

With the help of technology advancement presentation of old pieces of information into much more interactive, memorable ways, as this done with the help of VR and AR technologies. And because of this thing students get excited and they also get motivated.

Apart from creativity and motivation, there are many other essential advantages of VR in education., With safe training conditions, learners can be engaged in an environment reflecting actual scenarios, even extreme or dangerous.

Working on high or fire safety VR training simulators are prime examples of the benefits of VR in education. Without the risk and fatal errors, it is considered as an optimal protected VE where learners are experiencing the results of their actions.

7.4.6. Production-Driven Businesses and Automotive Industry

Implementation of VR is mainly done to reduce the time cost of that industry, and VR in the automotive industry does the same. Examples of the newest automotive applications include vehicle design, manufacturing workstation optimization, assembly training.

It is said that for industrial automation and remote assistance development VR is a must. For remote supervision, real-time instructions, object manipulation it gives digital opportunities, and as well as it also enables enterprises to automate production lines or troubleshoot issues in real-time. It is said that these types of software can be deployed to multiple users.

This is will result in motivating remote employees and encourage them to work simultaneously. It is said that from remote locations, VR social platform is the platform which allows collaborating in VR. With the help of VR training, workers are quickly prepared for the new machinery.

7.4.7. VR and Architecture

For building up the model digital tech development is considered as the breakthrough. After generating the great experiences in real estate and architectural design has moved to the next level in terms of quality. It is said that with the help of VR, conveying of idea is much easier, as with blueprints and building layouts idea cannot be conveyed an idea in such a way.

Every real estate agent knows that how to support their customer. As, the best way is considered is front door to attic and this way will help the customer most to make a purchase. So, because of this reason they host VR home tours. This way would really help the buyer, as buyers can step into virtual home and can have a look in all the corners with a 360° view walking up spiral stairs and determining furnishings (Figure 7.4).

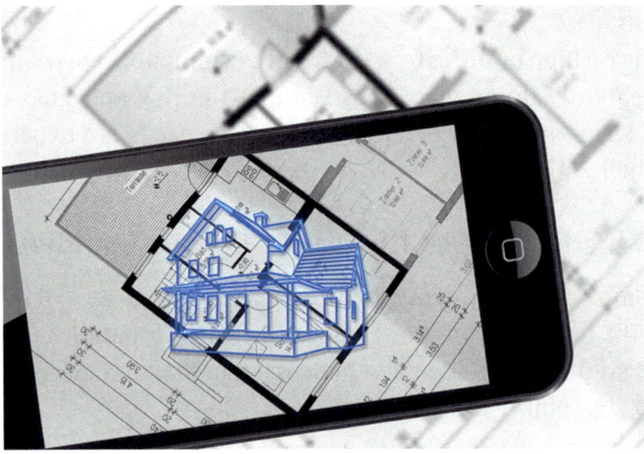

Figure 7.4: Application of VR in architecture.

Source: Image by Pixabay.

It is said that with the help of AR/VR-driven technologies not only the real estate tour is improved but also the interior and landscape design.

For exterior design applications, there are many other options of real-life to VR architectural visualization. There is an application by the name VR Gardens App and this app offers an interactive planner. It helps the users to design and buy their outdoor dream space via digital design tools.

7.4.8. Sports

It has been seen that VR has taken hold for years in the sports sector. Esports with the VR training and the consoles are the best example to remember. Just remember the consoles and eSports with VR sports training.

There are many sports bran but only the renowned sports brands allow sports fans to try new digital experiences. The best example for this is the Juventus VR app, which was created by Juventus F. C., in collaboration with Samsung that allows standing next to Ronaldo while training, celebrating winning in the locker room, or watching goal moment as close to the football field.

This is only done for fun. In the practical world, VR sports training is checked like racing with motion, force feed, manual shifters just like a real car, skydive, bungee jumping, skiing. VR simulators help the user to check the mountain slopes or climbing of wall.

7.5. BENEFITS OF USING VIRTUAL REALITY (VR) IN TALENT DEVELOPMENT

There are many technological advancements but out of all, the one advancement which is shaping today's modern workplace is the concept of utilizing VR in talent development.

Effective business training and development is done with the help of VR because of many recent advances in its core technologies, and VR usage extends far beyond the world of gaming. It is said that there are five benefits of integrating VR into a company's corporate learning experience and to increase efficiency, confidence, and skills.

7.5.1. Ability to Simulate Real Environments

A look into the training environment is very important because it will represent the clear picture. As the basic practice in a training environment is that employees are sitting in the conference room with a powerpoint presentation displayed on a screen (Figure 7.5).

Figure 7.5: Benefit of AR: Ability to simulate real environments.

Source: Image by Wikipedia.

So, the learning capacity is restricted for employees because basic presentation will not present a clear picture of the real world and the way of making them learn is not so encouraging. So, if the VR integrated in the business than it will provide chance to employees to learn via practical experiences, like a modern improvement on roleplay-based training.

For being one of the most effective ways of acquiring new knowledge and skills, experiential learning is a must.

7.5.2. Strengthened Engagement Levels

From the employee perspective, corporate education is considered as the negative light, along with the words boring and waste of time. When in management training VR technology is used, it enhances staff ability and encourage them to step in real scenario.

And with real scenario, it means that allowing participants to truly be there and interact with the VE. This will also motivate the learners, and they will not be tempted by distractions, which will also give them the opportunity to participate in the event.

7.5.3. An Increase in Knowledge Retention

It has been seen that knowledge retention has increased with the help of VR because in many cases it can cause employees to form an emotional connection with the material being taught. It will also result in boosting engagement. This will also give an opportunity to employees to practice the situations, and they are also having the freedom to do mistake without any negative impact on the company.

7.5.4. Reduced Training and Development Costs

Earlier when the company use to practice traditional training setting, every company rely on their dedicated trainers, new material, and training environment with the time blocked on every employee calendars.

Which will result in ongoing expenses along with productivity losses and inconvenience for employees. VR is considered as the one-time cost without resulting any overheads, and it can be done at employee's convenience, or even at home for remote teams.

Travel cost of business can also be reduced by implementing the VR solution for talent development. As with the help of VR, there is no need to bring the employees together from for training session different location.

This will result in a reduction in the expenses like airfare, transportation, hotel fees, food stipends, and now more can be spend on marketing, advertising, and promotion.

7.5.5. Exposure to Collaborative Scenarios

Effective training does not come from the solo experience. Watching video tutorials or reading a manual does not provide an adequate insight into how the workers should interact with others during the job. VR simulated environment help the employees or learners to become comfortable in speaking or working with others. So, with this employee will get chance personally to interact with clients and other workers as opposed to reading a script or watching a pre-recorded video (Figure 7.6).

Figure 7.6: Augmented reality gives exposure to the collaborative scenarios.

Source: Image by Unsplash.

It is said that without using the cost of raw material, anyone can update, customize, and adjust the corporate training with the help of VR technology. In talent development, the use of VR helps to keep the new trends and changing needs in an efficient and cost-effective way.

7.6. BENEFITS OF VIRTUAL REALITY (VR) IN HEALTHCARE

For improving the life of people, traditional medical approaches don't work well, so doctors are exploring more and more medical and healthcare organizations, and doctors are exploring VR proves to work better than traditional medical approaches.

There are some world top technology companies like Google, Microsoft, Sony, Apple, and Nvidia, who invest huge resources into the development of VR devices and applications in the medical industry.

There are some more useful and pioneer ways VR technology which is used in healthcare, and it includes the following subsections.

7.6.1. Medical Training and Education Using Virtual Reality (VR) Technologies

Education process and the medical training are the areas in which VR adds more color. General quality of medicine is improved by it. This app also has an additional feature of educating future doctors and also help them to learn better human anatomy, practice operations, and tech infection control.

Precision Genomics VR, Anatomy VR, and Airway EX Virtual Surgery Simulator are some examples of VR-inclined devices and platforms which it includes.

The quality of medical education and the learning experiences with the use of devices have gotten better.

As per the norm, there is a limit to the number of students at one time who attend an operation and learn the process. There is no other option for student to study while looking over the shoulder of a surgeon as there is no separate room for every student. The learning and teaching experience in medicine has moved to a lot higher and advanced level with VR.

It is a very great innovation and with the help of this now surgeon can stream their operations using a VR camera. And also, the medical students get the opportunity to have a closer look at the surgery process and get a better experience while watching and learning in virtual worlds.

Every detail from the start to the last move can be followed with their VR goggles. With this VR googles family members will also get an option to participate in the operation if they want to.

There is one more application of VR in medical education is that it can even help physicians to experience life as an elder person that is also applied in education.

It also helps the young doctors to reconcile the experiences of the seniors, as how they have taken care of the old patients. Sometimes they don't get to feel what their patients are feeling, so because of this, they treat them in a wrong way.

It is always difficult for the young doctors to recognize symptoms and problem, due to the non-experiential feel of old age and the illness come with it.

There is another VR tool that works like seniors by the name Alfred. It was created by Embodied Labs. In this every person can experience the life from the older patient perspective, but only for 7 minutes.

For medical students and young specialists, such a VR application provides a valuable experience in their study process.

7.6.2. VR-Based Mobile App for Healthcare

For solving a lot of medical problems, there is VR-based mobile applications that help in the treatment of different diseases. There are many other different medical apps which plays the important role in the healthcare like virtual medical games that help to get rid of pain and phobias, speed up rehabilitation, help patients in understanding the medical process and make medical study easier for students too.

It is said that like surgery simulator VR apps acts, and also people with disabilities VR app perform meditation. Other interested stakeholder can also use the VR education app, as it is not limited to use by students alone.

For example, there is a college by the name Royal College of Surgeons in Ireland (RCSI) who created a VR mobile app that makes medical education available to anyone even without a medical background. For non-medical users, the RCSI app has a particular model. In this application shows all the detailed information of every single medical procedures.

7.6.3. Pain Management and Relaxation with Virtual Reality (VR) Application

Brain pain is something that is unbearable by the patients, but the medical VR, which is also called VR therapy has proved useful in helping the brain process pain and even reduce the pains of patients. Patients with some chronic diseases' VR experiences act as a treatment.

The University of Washington HIT Lab conducted a study in which they have proved that the patients that engaged in VR felt less pain when compared to those who didn't and they were able to relax more.

This study explains every patient feel different pain, and the pain is depending on what the patient is thinking and doing at that moment. So, this means pain perception has a strong psychological component.

Getting inside the VE is possible only by interacting with VR people. For this process an individual required a lot of attention and brain resources too. Feeling of pain is less while spending all the attention on interacting with the VR world, so this means there is less focus on pain signals. Therefore, to get rid of chronic health issues such VR apps can help patients better than traditional treatment in the hospital.

There is one more application which is developed by the students from Simon Fraser University by the name Farm app. This app is mainly used for cancer patients, as it is a VR video game for cancer patients.

The main purpose of this app is to help teen cancer patients get distracted from the pain during chemotherapy treatments. For some time, this app takes patients into the imaginary world which help them to forget about the illness during that time, so there will be less pressure of pain and stress that they feel. It gives the different experience of life and circumstances to them.

There are many activities in the application which will make them happy and motivate them. Like, they can now escape the four walls of the hospital and have experiences of having to swim together with whales in a beautiful ocean, join helicopter rides over wonderful landscapes in the Poles, or get involved in games.

People suffering with the terminal disease are living their last days, so this application could also be useful to these patients. This will provide them good last days filled with fun. So, without even visiting the hospital VR devices help the patient release stress.

7.6.4. Physical Therapy and Rehabilitation with VR

To make the therapy process engaging VR technology allows physical and neurology therapists and it is considered as more effective than usual exercises.

During rehabilitation, there is research conducted to know the number of exercise patients do, and the result is that only patients do only 30% of the needed exercises. So, with the invention of VR technologies combining of physical exercise into virtual games for every patient become easier. VR physical therapy system developed by HTC Vive is considered as a great example of such a tool.

7.7. VIRTUAL REALITY (VR) FOR TOURISM

VR allows the viewer to explore all around the entire scene, as it refers to interactive images or videos. Regular image or video usually shot from a fixed viewpoint, but VR production captures every part of the location.

VR is used in every industry and in the travel industry, it is used to capture tourism destinations in a unique and immersive way.

With the use of specialist cameras, rigs, and software, this can be achieved. VR headset or a regular computer or mobile device on which the finished content can be viewed.

It is assumed by many people that the content of VR can only be viewed on a specialist VR headset, but this is not the case. Although VR is more immersive when viewed in this way, it can also be viewed on any device, including mobiles.

7.7.1. VR in Tourism Marketing

In the tourism industry, VR is utilized for marketing and is considered as the most common way. It is considered as the most powerful marketing tool, and it also being able to capture tourism destinations in such a memorable and immersive way. The feeling of being there is one of the greatest strengths which is allowed by the VR to the user.

7.7.2. 360 VR Tourism

The gaming industry is the one in which VR is started, so because of this, many people think of VR they think of it as CGI (computer-generated imagery). However, 360 VR, or 360 VR video are the different form of virtual reality.

The main focus of 360 VR is on the real world rather than CGI. So, in the tourism industry, it is important to show users a real location rather than a simulation. That is why it is best for the tourism industry (Figure 7.7).

Figure 7.7: A 360° image taken with the help of VR.

Source: Image by Flickr.

In a similar way, 360VR content in tourism is captured to regular image and video content. In order to shoot the scene a 360 VR company, like us at Immersion VR, arrives at the location with specialist equipment. The footage is then taken back to the studio where it is produced into VR content using specialist software.

There is some industry in which 360 VR is commonly used that is education, real estate and also in online marketing. In the tourism industry, it is something in specialized form.

7.7.3. VR Technology in Tourism

In the tourism industry, VR can be used in many different ways. As with the expansion of technology the VR within the tourism is also improving. The technology is evolving at a rapid rate. VR video and VR photography are the main VR technologies that are used in the travel industry.

7.7.3.1. VR Tourism Videos

Just like a standard video, a VR tourism video works the same. For this particular video, it is available on various social platforms and websites. Apart from this, the operator can sightsee the entire scene at the same time the video is getting played.

Through omnidirectional cameras, VR tourism videos are taken which comes under specialist cameras. Different angles can be captured through these cameras in one time. Once the film is being captured, it is taken to the studio for compilation and editing in order to get the final product for VR tourism video.

There are two types of VR tourism videos:

- Monoscopic VR tourism video; and
- Stereoscopic VR tourism video.

For everyday devices such as computers and mobiles Monoscopic VR videos are used as it can be easily viewed on these devices. The user can easily move it or rotate it around in order to view the entire scene. Viewer can turn his/her head to view from different angles to sightsee any scenes.

Unlike Monoscopic, the Stereoscopic VR videos does not work on everyday devices. They are only made for VR headsets. Even though they are more expensive as compared to Monoscopic VR videos but still they provide with enveloping traveling experience for travelers.

To explore the surroundings in a realistic way, these videos feature head tracking so the user can move their head.

7.7.4. VR Tourism Photography

It is said that production of 360 images of the travel destination is involved by the VR tourism photography. To be viewed on regular device such as mobile and desktop, these images are designed. Virtual tourism photography works much like a VR tourism video but with still images.

With the state of the art, the images are taken with DSLR cameras on specialist rigs which allow for the capture of 360°. As compared to VR tourism video DSLR cameras are used because it allows for higher resolution images.

Just like the regular images, 360 images uploaded to social media and websites for the user to view just as easily as regular images. As these images are not enhancing as VR videos but they are cost-effective and easy to produce.

To explore the hotel completely is done with the help of 360 photography which allow the user to explore a hotel and its surroundings in an enhancing and interactive way.

7.7.5. Applications of Virtual Reality (VR) in Tourism

Applications of VR in tourism include:

- Virtual reality travel experiences;
- VR tourism content for social media/websites;
- Virtual hotel tours.
  - ***Virtual Reality Travel Experiences:*** VR tourism videos made with the VR travel experience and it is made for VR headsets. With the aim of creating a feeling of present at real destination, these virtual travel experiences are used.

It is beneficial for the user, as it provides unique and memorable experiences to the user. There many travel companies and travel agency which are accepting this technology, as they feel that this technology is continually growing and they promise a bright future within the industry.

- ***VR Headsets in The Travel Industry:*** For the user VR headsets provide the most realistic VR travel experience. A special software is used by the VR headset, which tracks the movement of the user's head. In real life, this will allow the user to explore travel destination.

Currently, the number of people that own a VR headset is rising at a fast rate. This growth in headsets can largely be attributed to the gaming market, where the technology is being pushed hard.

In the VR headset, there are many online platforms investing including Google, Facebook, and Amazon and VR content, promising a bright future for this space.

In order to create a spatial audio and stereoscopic content, VR headsets becomes more expensive for adding these special qualities. Moreover, for traveling brands, this cost becomes worthwhile to compete in this global world and to standout from other companies and also to give their clients never-ending experience.

7.8. BENEFITS OF VIRTUAL REALITY (VR)

The benefits of VR in tourism include:

- At a travel destination user are allowed to imagine;
- Being able to showcase 360° of a destination in high resolution;
- Allowing the user to explore a scene at their own will;
- Creating unique experiences and memorable for the user;
- Creating unique brand engagement;
- Allowing travel companies to stand out from the crowd;
- Providing travel experiences to those that cannot travel;
- Reducing impact of tourism on vulnerable destinations.
- **VR Tourism Statistics:** A research conducted in Germany by Statista almost 50% of people would use VR as a tool for choosing their holiday destination (providing it was free). 13% of those surveyed were actually willing to pay for the VR (Figure 7.8).

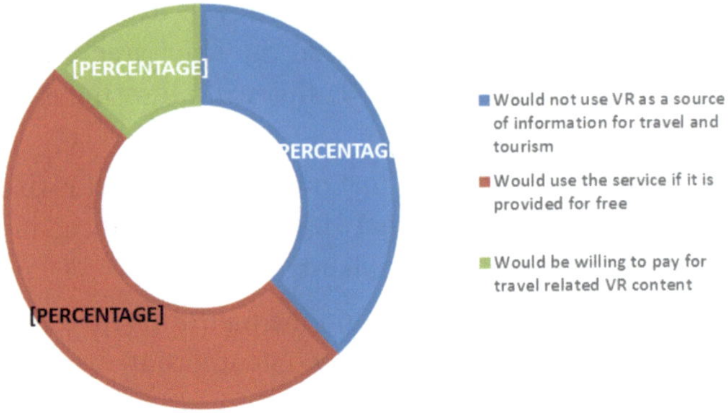

Figure 7.8: Statistics of VR in tourism.

Moreover, a research is carried out by the search carried out by Tourism Australia found that almost 20% of consumers had used VR to select a holiday destination. Around 25% of consumers said they planned to use VR in the future to help them decide on a holiday destination.

7.8.1. Travelers Can Get a 3D Tour Before Touching Down

The experience of knowing where one can land at a sight is unspeakable in a way. The best part about the 3D tour is that the tourist can experience their destination even before arriving through this VR.

Unlike pictures, this 3D experience allows users to become part of the reality just like in the real world. This experience allow tourist to feel different cultures around the world of any city without actually setting foot in that country and it makes them feel more like home rather than outsiders when they reach there.

7.8.2. Virtual Reality (VR) Can Help with Planning Sights to See

In the real world, tourist are always short in time, and they are unable to experience everything when they go out for vacation. Some of the other things will always be left behind due to lack of planning. The tourist always misses some of the major sightseeing places, but through VR, they can easily sightsee all important places and can learn about different locations and their names before time.

This experience also allow user to decide whether they really want to experience the same in real-world and to make relevant choices for future vacation planning. Due to this technology, users get to entertain themselves with locations that are more interesting to them in real-time because of this VR sightseeing experience.

7.8.3. Travelers Can Use VR Technology When There Are Delays in the Travel Schedule

VR technology also help user to spend their free time when their flights are delayed at the airport, due to unspecified reason and one has to wait for numerous hours at the airport. VR technology become the part of their amusement and makes them excited again to travel the spaces while waiting for the trip ahead.

7.8.4. Virtual Reality (VR) Is Great for Touring the Plane Ahead of Time

VR 3D reality tour makes people aware about the safety within the plain such as emergency exit, safety measure to be taken care of in advance even before taking the flight.

7.8.5. Travel Coordinators Can Create Engaging Presentations Using VR

Travel coordinators can consider the benefits of VR in the form of presentations. They can form potential clients by creating a VR tours of different cities. Imagine if anyone can take a tour before planning a trip ahead and before paying for the same. It helps customers to become aware of their locations and know they will be enjoying at their destination. It helps them in making more improved planning.

7.8.6. Travelers Can Learn More about a Specific Location before Touring the Sight

Walking into the history is one thing and just listening and reading about it is another. VR comes as an amazing experience where they can walk into the history location where they can gain actual experience of major events happened in the past. By gaining the knowledge beforehand, it helps tourists to visit the same spaces with their complete knowledge. VR gives them more knowledge about places unlike any tourist guides or history session.

7.8.7. VR Is a Great Way to Advertise on Social Media

Travel companies can be benefited through VR technology by giving their customer a 3D experience before booking, unlike mere pictures being posted over social media. Travel companies can increase their business through this technology and can help them beat their competitors.

7.8.8. Virtual Reality (VR) Can Help the Traveler Find Locations on the Map Faster

In order to make more efficient plans and to locate nearby necessary locations such as a bakery or a coffee shop VR can help the same without losing much time and can help users in finding timesaving locations just 10–30 minutes away.

7.8.9. VR Is a Good Way for Travelers to Plan How Much They Need to Pack

VR also help tourist in choosing the hotels. In today's times where there are so many fake advertisements regarding the hotel booking, misleading visitor's VR helps tourist to plan and take a tour of rooms and hotels for a better booking experience. It makes them aware about room sizes, hotel rating, closets of the room, so that they can bring the stuff and luggage accordingly.

7.8.10. Virtual Reality (VR) Is Useful When Travelers Must Choose Between Destinations

VR helps the family who cannot afford vacation to travel to different cities. It also helps travelers to choose the location for their destination by comparing different sites through VR sightseeing and by exploring different sounds.

7.9. VIRTUAL REALITY (VR) AND ITS IMPACT ON BUSINESS

> Virtual reality is especially effective for marketing, because it creates an opportunity for businesses to establish a strong emotional connection among target consumers and their product.

For describing AR and VR immersive technology is the catch-all term, and it is now having its moment and after years of development and improvements has finally become mainstream.

There are major tech brands such as Google, Samsung, HTC, etc., which are launching their own versions of VR devices, it is said that the popularity of these devices and virtual content is sky-high. In different industries and domains, VR has already seen a wide range of applications.

VR is the thing which is not limited to the single industry or a single domain or vertical. Although the gaming industry seems to be the key beneficiary of this technology. There are various industry sectors such as healthcare, travel, business, etc., in which VR has already created a storm.

So, after this a question which comes to every individual mind that how can VR influence businesses? So, in this chapter, the main business benefits of VR and what impact it can have is listed.

7.9.1. What Will Be the Impact of Virtual Reality (VR) on Business?

It has been seen that VR is one of those technology which has a real business value as well, as people often relate VR to gaming and headsets. In the today's world to incorporate VR into their routine business processes there are more and more businesses which are coming with newer and innovative techniques and they are also trying to make their operations more productive and efficient. There are some of the major business benefits of VR and how it will have an impact on the business are discussed herein subsections.

7.9.1.1. An All-New Level of Product Prototyping

With the help of VR, companies will get the benefit to better visualize and design the product that they are developing like the product which has been never been developed. Earlier it was not possible to have the design of the product, but with the start of VR, the product being developed can be precisely designed, analyzed for its functionality.

And there is one more benefit that it can be modified multiple times if required before it is sent to production. As well as, it will also give the benefit to the company's decision-makers and the product's end-users as with the use of VR they will also be able to give some constructive feedback about the product being developed, which in future can be incorporated into the product in the initial stages of development.

Therefore, this way is considered as a great way to quickly detect design problems and deal with it sooner, which will also help in avoiding post-production complications.

7.9.1.2. Reduced Business Travel and Efficient Business Meetings

Any business can be revolutionized with the help of technology which reduces the travel and communication gap. Earlier, there was a very high cost associated with the business travel because of which distributed teams was considered to be highly inefficient. But this gap has reduced a lot, with the development of the advanced technologies and infrastructure.

It is also said that in future VR will play an important role in future, as it will further reduce this gap, and in the future, it will also help businesses in successfully conducting virtual meetings involving different teams spread across the globe.

7.9.1.3. E-Commerce Advertising Will See a New Side

There are some obvious industry sectors like gaming and movies where VR is expected to thrive, and apart from these obvious industry sectors, e-commerce will be one field in which companies can gather huge business benefits of VR. However, to draw maximum benefits from them, e-commerce has some uncharted regions which need to be identified by companies.

And this will be possible by using VR, as it is an important part of the purchasing cycle, and it also creates a whole new way in which people can shop on e-commerce websites.

7.9.1.4. VR Can Provide Competitive Advantage to Businesses

Creating a virtual store is more convenient, and it can also be created very quickly in relation to a traditional product launch or physically building a new store because it is considered much easier as compared to physically setting up a store.

By using this technology, the speed of marketing will increase and along with the speed, the store can be planned or build in complete secrecy. Building store in secrecy will helps businesses to stay ahead of the competition.

7.9.1.5. Interviewing Candidates Will Be Easier

With the use of VR technology taking interviews of the new applicants for the job than it will be much easier for the human resources department in the future. It will help in a way like from the virtual conference room they will be able to interview candidates from different locations face-to-face.

And it will also help them to analyze and view the candidate's responses and body language. There many options available for the large companies that they can even have remote offices for candidates to use the infrastructure and participate in the interview or discussions.

7.9.1.6. The Future of Retail Will Be Very Different

It has been seen that to identify various cultural, consumption, and social trends of the customers through VR, companies will be able to invest in extensive research and development, and it will also be able to obtain accurate industry forecasts.

Therefore, it is said that VR will have a huge effect on the future of retail, as it will also help the businesses to understand consumer behavior, trends, and product consumption patterns in different scenarios.

7.9.1.7. Virtual Conferences and Meetings Will Be Popular

There are many teams in the company which are spread at different places, and it is very difficult to gather all the participants and have meetings for teams that are spread across the globe. There will be a high level of misunderstanding and lack of communication, if there are fewer meetings between the employees, which can have an adverse effect on the business. And surely meetings and conferences will be much more interactive and fun with VR technology.

7.9.1.8. VR will be an Important Tool in Training

It is said that in the coming decade professional business trainers are expected to use VR extensively. There many good features which are associated with the VR but the most different feature of VR allows the participants to look around in the room and interact with those who are online.

Along with this different feature trainer can record the training sessions and trainers will also be able to conduct live training sessions and seminars, which will help in making these training sessions more interactive and bouts.

7.9.1.9. Affordable Virtual Designing of Structures to Visualize the Project Better

It has been seen that VR will be more beneficial for the engineers, architects, and other professionals who are into architectural planning because it will allow them to clearly visualize the structures they are designing.

A lot of money and time will be saved; otherwise, the money would be invested in making smaller models or 2-D renders. Therefore, by using immersive 3-dimensional technology to design structures, businesses will be able to achieve massive growth through VR.

7.9.1.10. Offering Virtual Tours Will Be Possible

There are many uses of the VR, but the most important use of the VR in business is the ability to provide virtual tours of different locations. This will help in many industries, but it will be more beneficial in the real estate industry where the real estate agent or owner can provide a virtual tour of the property to their clients.

This will be also beneficial for the clients, as it will help them to visualize the architectural plan better to make an informed decision. Customer can get a 3-dimensional view of the house they are planning to buy or rent by using the VR.

7.10. CONCLUSION

Overall, the VR and AR come as a great tool for future technology and the increase usage of the same could lead the future generation with better experiences in every field. In the education system, it could not only guide future students for exploring and reaching new heights in terms of science and technology but also could help them to increase their knowledge in a better way and to compete with fast-paced environment. Moreover, in health as well as tourism industry the introduction of VR can lead to new experiences and can help customers to get better experiences.

REFERENCES

1. Business World iT, (2019). *Ten Uses and Benefits of Virtual Reality in Tourism | Benefits of VR*. [Online] Available at: https://www.businessworldit.com/ar-vr-technologies/10-uses-and-benefits-of-virtual-reality-in-tourism/ (accessed on 29 April 2021).
2. Concannon Business Consulting, (n.d.). *Five Benefits of Using Virtual Reality in Talent Development*. [Online] Available at: https://www.concannonbc.com/virtual-reality-vr-talent-development/ (accessed on 29 April 2021).
3. Flatworldsolutions.com. (n.d.). *Virtual Reality and its Impact on Business-Flatworld Solutions*. [Online] Available at: https://www.flatworldsolutions.com/IT-services/articles/virtual-reality-impact-on-business.php (accessed on 29 April 2021).
4. Horiachko, A., (n.d.). *Nine Ways to Use Virtual Reality in Healthcare*. [Online] ReferralMD. Available at: https://getreferralmd.com/2019/09/9-ways-to-use-virtual-reality-in- (accessed on 29 April 2021).
5. Immersion VR. (n.d.). *VR For Tourism | The Future of The Travel Industry | Immersion VR*. [Online] Available at: https://immersionvr.co.uk/about-360vr/vr-for-tourism/ (accessed on 29 April 2021).
6. Soeonline.american.edu. (2019). *Virtual Reality in Education: Benefits, Tools, and Resources*. [Online] Available at: https://soeonline.american.edu/blog/benefits-of-virtual-reality-in-education (accessed on 29 April 2021).
7. Web and Mobile Apps Development, UI/UX Design | OpenGeeksLab. (n.d.). *Eight Advantages of Virtual Reality in Business*. [Online] Available at: https://opengeekslab.com/blog/advantages-virtual-reality-business/ (accessed on 29 April 2021).

CHAPTER 8

Future of Virtual Reality

CONTENTS

VR craze has been increasing at a rapid rate, and in the future, with advancement, it is expected that VR will create its space in more arena. In the chapter, the future of VR, the meaning of virtual reality (VR) is discussed. The chapter also tries to explain whether, in future, VR will become the part of future corporate training. It also highlighted that VR is making it ready for future form of communication.

Some of the important considerations for enterprise immersive learning are also demonstrated in this chapter. It also shed some light on future challenges in VR. Some applications of VR are also highlighted in this chapter. The role of VR in socializing is also explained. It also discussed the role of VR to change the future of music. Trends and opportunities of AR and VR in the future of payments is also discussed. It also explains the ways in which VR can impact the future of tech careers. At the end, the chapter highlights some key players in this field.

8.1. INTRODUCTION

It is well known that virtual reality (VR) is in existence for some time now, with the first gloves and goggles being developed in the mid-1980s by Jaron Lanier, founder of VPL Research. Almost 40 years later and this technology has become combined into the mainstream and is now something that can be access and use by the general public. Majority of the people now enabled with the opportunity to use VR or see it in action, whether it's been at a VR center, a theme park, or having a turn on a friend's headset.

Whilst VR is widely popular for transforming the entertainment and gaming industry in specific, with handhelds and consoles adapting their games for VR, the gaming experience has been totally renovated and enhanced all around. The connection between player and game becomes far closer as gamers are entirely immersed into their own gaming world.

Apart from the gaming and entertainment industry, VR has also started to transform other industries as well such as retail, healthcare, and education, which is already assisting tasks in the sectors, from facilitating patients with nervousness relax, and aiding students to enhance their knowledge base and giving them real-life examples.

But what are the most exciting changes that one can expect from VR in the next few years?

It is expected in the coming few years, there will be a significant transformation from a simple form of sunglasses to experiencing what life would really be on Mars. In addition to it, it will also assist people to overcome some of their biggest fears and phobias. VR will likely to do far more than what we can expect from it in the future.

In the next two years, it can be expected that the technology will be able to make an experience for individuals where they would not be able to identify the difference between the real and virtual world, with the integration of VR and augmented reality (AR).

Michael Abrash, the Oculus chief scientists who made this forecast states the problems in reconstructing the real world, and demonstrates that augmented VR will be an essential part of VR. He also states that doing merger of these innovations will see a technology that can be used for a longer period and for a variety of purposes than it can be today.

Whereas, at the same time, there is a need to do more work, Abrash is self-assured that a technology will arise where the boundary between real life and VR will be far more unclear and blurred.'

In the later future, one can witness the innovation in the VR that are beyond the thinking of normal persons. As it is well known that we are presently living in the midst of a space race to Mars among some of the world's most eminent billionaires such as Richard Branson, Elon Musk, and Jeff Bezos, who all have made their mind to take humans into space first and eventually reaching Mars. But with ticket prices reaching exorbitant amounts, the mainstream of us won't be taking one of these flights.

Fortunately for all of us, it is the VR that can provide the similar experiences which all of us would love, making all the to see what life on Mars is really like. Not only is this a feasible and viable substitute, for those of us that do not have enough money to afford a real ticket to this red planet, but it also works best in the people working in the space industry wanting to use it for training purposes.

VR's assistance to experience life on Mars will also allow research and development to take place at a more rapid rate and in a sustainable manner and this assisting in accomplishing human missions.

8.2. WHAT IS VIRTUAL REALITY (VR)?

VR can be defined as a group of technologies that assists people to interact efficiently with 3D computerized databases on a real time basis using their skills and natural senses. This definition circumvents any reference to a requirement for head affixed displays and instrumented clothing such as suits or gloves, as was the bias in the early 1990s.

Although this still now known by the name immersive technology and is evident today, only 10% of VR applications warrant its use. The main strength of VR, be it in training or design, is that it augments and supports real time interaction on the part of the user (Figure 8.1).

Figure 8.1: Unique experience in VR.

Source: Image by Pinterest.

8.3. IS VR THE FUTURE OF CORPORATE TRAINING?

Soldiers, astronauts, and surgeons have been coached for several in VR. People usually learn faster by doing one thing again and again and by getting feedback when they make any error, which is why these high-stakes lines of work are natural applications of the medium (Figure 8.2).

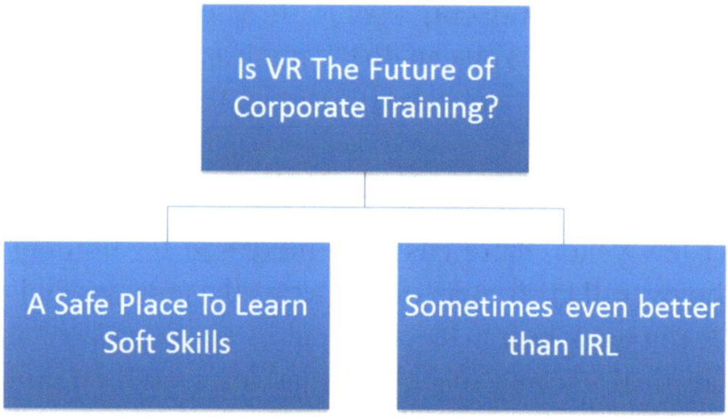

Figure 8.2: Is VR the future of corporate training?.

But over the last some years, the expense to use VR has plunged, and the technology has spread its feet into more general use at Fortune 500 corporations, where employees working in industries such as logistics, retail, and customer service are performing in VR headsets to make their jobs better.

A more efficient way to learn procedures. It is well known that there are hundreds of academic studies on procedural training, and the literature is highly developed to embrace a number of meta-analyzes—for instance, comparing VR surgical training methods with other techniques.

The VR training is almost work equivalence to face-to-face training, in addition to the economic savings by using VR. Apart from this, when people are trained with VR, the total time expended to train people is comparatively less than other conventional techniques.

For example, recently, Walmart has given training to more than millions of its employees with the help of VR. One of the most commonly used modules is "The Pickup Tower,"—which is principally a large kiosk that allows customers to pick up online orders. Trainees has been given step-by-step instruction on how to run this new machine, with instant feedback if they do any error.

Before introduction of VR, each person spent the majority of their day on training inside expressly designated stores, with some e-learning and some hands-on training. With the help of VR, the total training time has been reduced from 8 hours to just 15 minutes, with no drop in efficacy. Given the fact that all Walmart associates nationwide would be required to train on The Pickup Tower, VR should return over a million full days of work.

To quote Heather Durtschi, senior director of content design and development at Walmart, "You can do the math as to what the savings would be."

8.3.1. A Safe Place to Learn Soft Skills

Over the past few years, there has been increased for training "soft skills" to enhance the overall customer service as well as managerial skills. It was found that VR can work best in teaching public speaking. As it is well known that giving physical training of public speaking would be quite expensive, VR is a game-changer in terms of both cost and ease (Figure 8.3).

Figure 8.3: VR in corporate training.

Source: Image by eLearning Industry.

VR provides an exceptional balance across experiments—it is immersive enough to make people stay motivated while they are taking training. In addition to this, it also provides a safe environment where learners are less self-conscious about speaking freely in comparison to talking to real people. Thus, it has a better rate of improvement as compared to physical training class.

For instance, Verizon has constructed and executed a module to train employees working in call-center on how to de-escalate a conversation with customers who are dissatisfied. Trainees are given a chance to practice speaking and active listening as a customer conversation becomes gradually tense.

According to internal record gathered by Verizon, it was found that because of VR, there has been a significant increase in the effectiveness and consistency of the training, in addition to the time needed to train employees to just 30 minutes from 10 hours per person.

> According to Cleo Scott, Director of Global L&D for Verizon Business Services, "As they went back to work and we tracked their progress through the supervisors, the employees were much more confident, because they were more aware of themselves in how they were handling the customer."

8.3.2. Sometimes Even Better Than IRL

One of the greater challenges for companies, especially during the situation like COVID-19, is to integrate new employees into the cultural norms of the company. With the help of VR, one can try to bridge the gap between different personality and establish common culture without academic precedent.

Sprouts Farmers Market, which is the multinational supermarket chain, relies heavily on employee culture to differentiate their brand, concentrating on core values such as "Embrace Healthy Living," and "Respect and Serve One Another." As Sprouts endures to set up new stores and hire across the country, they need to onboard new employees.

Sprouts formed a standard of VR experiences aimed at demonstrating these core values—for instance, an employee might teach an anxious mother who just learned her son is sensitive to gluten about how to shop for entire new diet, or an employee may choose to transport a watermelon to an elderly sick customer who can driveway long to pick up his desired food.

Instead of tutoring explicit skills, they have applied what cognitive psychologists call an Exemplar Model, emphasizing a variety of salient examples which work in tandem to shed light on an abstract theme.

A group of around 300 employees were tested on this foundation stone, half of whom were trained with VR, and the rest were given training on powerpoint. Almost 48% of the trainees who trained on VR grasped all six concepts completely, in comparison to only 3% who used conventional methods. Cindy Chikahisa, VP of Store Operations for Sprouts summed it up: "I'm so grateful we launched this before COVID-19. I can't imagine hiring thousands of people—which we've done over the last few months—and trying to give them a great onboarding experience during a pandemic."

VR was increasing its footprints even before COVID-19, but the global pandemic which trigger the situation to work from home is fast-tracking the necessity for such tools. These three companies are mounting despite the crisis, and their need to train employees securely, and in an efficient and effective manner at scale grows as well. VR is the perfect medium for this moment.

8.4. FUTURE OF COMMUNICATION: VIRTUAL REALITY (VR) IS STEPPING UP

It is generally seen that the way in which we communicate with each other is an ever-transforming discipline. Each new technological advancement offers new opportunities and new channels for individuals to consume and disseminate data (Figure 8.4).

Figure 8.4: Virtual reality for communication.

Source: Image by innovate MedTech.

In the past few years, AR and VR have elatedly risen in eminence. From the employment of VR in corporate communications network to the historic achievement of games such as Pokémon Go, there are clear signs that we all need to get prepared for this advancement.

With the rise of coronavirus, a lot of workforce has been forced to work remotely and could at last usher in their usual utilization of AR and VR at home, or perhaps give the technology an impulsion on the way to become standard.

Digital platforms have some time early empowered a more innovative way to connect and the way with which online content is created is ever changing. VR, 360° immersive photography and video are revolutionizing how we tell stories. So, what's the importance that it carries here for the eventual future of communication?

With VR, individuals can present data as a multi-dimensional model on a 3D material instead of 2D Excel sheets, visual charts, and pie charts. Data turns into the settings that persons can stroll through.

Let's take the example of the VR network visualizer which was formed at the University of Cambridge's Institute for Manufacturing. The VR tool authorizes its customers to get the assess of supply chain network of their organization with the help of a VR headset and regulators.

The VR potential reach out past learning for the objective of individual enrichment. Take into consideration the economic implications with respect to acquiring skills or job training that could be applicable to real-life jobs.

Walmart is considering to implement VR technology into their organization with the ultimate goal to train their employees. From handling the holiday rush to tidying up passageways, putting up scenarios via VR is often indicated to as yielding more retention in comparison to paper or video training.

In addition, Google's Daydream VR group is presently exploring different options staring a virtual coffee machine to assists baristas understand the fundamentals of their responsibilities without getting their hands dirty.

Video conferencing has been essential in uplifting remote interactions between business experts and has seen astonishing achievement through platforms, such as FaceTime and Skype. Currently, VR is expected to transform that technology further.

Now, it is possible for two business experts positioning at two far sides of the world to 'meet' virtually in a similar space to have a face-to-face conversation. Besides joining with shake hands and getting a hand through dainty air, that association would be as real as they are present in the same room.

AltspaceVR wants such sort of VR chat meetings to be part of their daily life. With their acquaintance application for smartphones, they are diving VR talk similar to Skype meetings, yet it's more about mutual action than unadulterated discussion, with the substitute to 'meet' in conditions, for instance, labyrinths, game rooms, or virtual craftsmanship exhibitions.

It is easy to set up virtual settings, and it requires less efforts in comparison to standard video gatherings. United with their new digital appearance, members can feel freer to share innovative thoughts. Architects at Kia and Hyundai are as of now exploring VR options to design their self-driving cars system.

VR is now gradually transforming about how we learn and communicate with each other, though, with the time, we get to know when this technology will become the standard. Irrespective of whether through acquiring a totally new viewpoint or learning skills on the world, the opportunities are evidently endless for those who are ready to open their mind about the idea of virtual environments (VEs) and immersive technology.

With the Industry 4.0 revolution and increase in the adoption of digitization, there has been a dramatic shift in the skill requirements for the future workforce. According to research from World Economic Forum, it was estimated that by 2020, closely 35% of the top skills required among all job families will amend; hence, there is a growing need to emphasize on corporate training.

According to Statista in 2017, it was found that corporations are expected to spend almost $362.2 billion US dollars on corporate training initiatives across the world. Although, these sessions are mainly provided in conventional forms such as online training modules or classroom-based seminars.

While passive learning and memorization has the model used earlier, today's workforce needed a more dynamic and enduring approach to training in which employees' study through practical experience and real-life examples.

Moreover, it is imperative on the part of companies to make a realistic version of risky scenarios to test compliance and safety protocols-such as what actions should be taken if an event of fire occurs or how to work securely onsite in an unsafe area.

Experiential learning has been viewed for long as the most effective way to learn, and studies have proved that learning through experience and real-life examples has enhances learning quality and progresses retention rate by as much as 75%.

8.4.1. Benefits of Immersive Learning

- ***Mirror Real-Life Situations:*** Immersive learning is one of the effective ways in stressing things through visualization. By offering environments that are closely associated with real-life situations, employees can attain higher levels of expertise in minimal time.
- ***End of Distance:*** According to a survey conducted by Accenture Technology Vision 2018, it was found that 36% of executives identify eliminating distance barriers between people and information as a driver in their adoption of VR solutions. Through immersive experiences, businesses can attain expertise in hundreds of skills sitting at anywhere in the world. In addition to it, VR can also provide remote collaboration and remote guided tours.
- ***Reduced Operational Costs:*** It is generally seen that companies that adopt immersive learning can easily reduce their costs on employee travel as well as on transporting equipment to training locations. In addition to this, they can also even save space on real estate. The trainers can themselves be part of the VR programming, so organizations can decrease faculty costs.
- ***Learning Through Mistakes:*** One of the most imperative benefits of immersive learning is that people do not need to worry when they make any mistakes, which can be costly in the real—both in terms of safety and machinery. Training for dangerous environments, as well as simulations that permits people to practice presentations, reduce actions that do not back inclusion or that could negatively influence a client deal can all be succeeded through VR.
- ***Increased Engagement:*** With the potential to build in gamification, immersive learning can offer fun or joyful experience. When trainees are interested and engaged, it leads to enhanced retention.
- ***Better Analytics:*** It is usually seen that VR captures enriched user data—behavioral, heat maps, eye tracking, and gesture tracking. Management can evaluate immersive learning experiences and check results via automated reports, which assists position employees for growth and development in the future.

8.5. IMPORTANT CONSIDERATIONS FOR ENTERPRISE IMMERSIVE LEARNING

8.5.1. Identify Meaningful Immersive Use Cases

In order to make the optimal use of a corporate training budget, it is essential for organizations to select the right areas for VR-based learning. It is important to note that Immersive learning is more efficacious in use cases that entail a first-person perspective and test response and interactive practice.

It is primarily because of the fact that presence—the cognitive and emotional effect that can be attained by creating true immersion—leads to more effective retention and comprehension rates.

Although, some use cases can be properly effective by forming focus-based engagement and letting participants to solely look around. It is also significant to initiate a preliminary business cost/value benefit analysis of the use cases and to evaluate the minimum feasible product needed.

8.5.2. Design a Compelling User Experience

- *1. **Interface/Interaction Design:*** Constructing complex user interfaces (UI) that have a sharp learning curve will be counterproductive; rather than design UI that is simple and fun. If there is need of interactivity, it will be imperative to engage in the proper design expertise as inadequately designed use of controllers rapidly abolishes the positive effects of presence and can annoy learners.
- *2. **Fidelity Level:*** The use case should outline the range of visual fidelity the experience needs, often communicated in terms of head-mounted display (HMD) resolution and polygon counts. The trade-off on fidelity/ size of the 3D assets comprises cost, size, and effort of the content.

In terms of HMDs, first-class devices such as StarVR, with its bigger field of view (FOV) and higher resolution, delivers rich experience for use cases that needs extremely high resolution. While on the other side of the spectrum are the novel all-in-one devices that have lower processing power and resolution.

- *3. **Multi-Participant:*** It is important to note that certain learning scenarios would be more impactful through the backing of numerous participants being in the same VE. It is usually the case in the student-teacher model where the teacher is guiding a group of students via experience and real-life examples, potentially administering over "control" to a student at certain points so they can interact with certain entities.

The participants may be co-located (which comprises of extra considerations around technical and movement selections for tracking) or maybe in numerous locations (which comprises of technology and network speed considerations that backs such sort of user behavior synchronization).

- *4. **Physical Environment:*** For learning experiences that necessitate movement, it is essential to take into consideration the physical layout and configuration of the setup. Multifarious setups followed by tracking sensors on the wall may be satisfactory for a dominant location but could prove challenging to cope in a larger number of locations (for instance, national network of retail stores). In such scenario, a naiver 3° of freedom (DoF) experience may be more feasible and practical.

In the future, integrate touch: For specific use cases, learners would be able to get an in-hand experience of an object via haptic devices when they touch the object in VR. Haptic technology is basically known for offering a higher sense of presence and can be applicable at anything from training surgeons to creating more precise and immersive virtual games.

Current touch controllers offer comparatively naive haptic feedback such as vibration. Although, devices such as HaptX, are speedily refining, directing toward a future when true haptic will cover the whole body.

8.5.3. Build Analytics from the Start

With VR automated and analytics reports, companies can train, evaluate, and manage the soft and hard skill sets of employees to progress training effectiveness. Employee actions, as well as behavioral and emotional responses to distinctive situations, can be apprehended and assessed by incorporating VR with other technologies for gesture tracking, eye tracking, and voice recognition.

8.5.4. Choose Appropriate Technology

The most significant technology decisions center on what type of VR medium is most suitable or adequate one (VR or AR) for the use case, the HMD/device selection and use of controllers, platform manageability and integration requirement.

For formal learning, it is generally seen that VR is considered as the ideal medium given the level of immersion it generates. For learning in areas such as police force empathy or de-escalation training, creating presence is essential in order to make the desired impact.

AR works best in a scenario that needs interactions between people and objects in the real world. AR also considers as the best option when there are security or safety concerns about participants wearing full VR HMDs during training.

For simple training where focus-based engagement is adequate, companies can use more economical and easily managed three DoF mobile VR solutions such as Oculus Go. Training that needed full movement and interactivity to create the preferred effect (e.g., conservation on complex machinery) needs a PC-driven device such as Oculus Rift or HTV Vive.

Although, with new developments in all-in-one devices emerging in the market as early as 2022, it may be probable to offer experiences that are presently only exists on PC-driven devices. As it is discussed above, certain premium VR HMDs may be adequate if the visual fidelity quality level must be enormously excessive.

Presently, it is usually seen that the majority of AR experiences run on HoloLens or smart phones. In the next few years, AR smart glasses from companies like Apple and Magic Leap, along with advancements in HoloLens projected in 2022, will unravel more AR immersive learning prospects.

The choice to incorporate with other technologies such as the internet of things IoT and artificial intelligence (AI) varies on the sort of end-user experience needed and is commanded by the use case and benefits/ cost considerations.

If adequate, it is also imperative to ponder and design for integrations with other corporate applications such as user authorization/ identification platforms and learning management systems.

As enterprises are consistently expanding their use of immersive learning, they will require to take into consideration other technology decisions around how to best manage and scale the program. Instances include device management solutions, cloud-based rendering infrastructures, network design, and security management.

Network speed and bandwidth are especially imperative if the use case includes multi-participant necessities. Working with an experienced enterprise planned with proficiency in VR technologies will be essential as these decisions are formed.

8.5.5. Establish Governance Board

It is often seen that immersive learning initiatives would be more fruitful if directed by a strong governance board that is tasked with crucial responsibilities such as:

- Looking for the success metrics;
- Identify and arrange supplementary learning modules and implement effective ones across the organization;
- Maintain general technology architecture idea, integrating new technologies as adequate;
- Ensure governance of third-party content;
- Make 3D apps and content easy to access via a repository or enterprise store to promote reuse;
- Describe deployment criteria that comprehend standard physical environment configurations, use of multi-participant models, etc.;
- Outline and impose ethical standards around using VR given the power effect of presence;
- Collection of data and create measurement techniques for success metrics;
- Measure overall employee capability and try to look for the ways that leads to training improvements.

8.5.6. Perception Requirements

Our perception of the real world is constructed and derived by the accumulation of information from our various sensory organs. VR is basically concerned with giving us this perception through some or all of these senses; this can be attained via feedback one obtains from sensors. The extensive categories of the perceptions accomplished through VR are:

1. ***Visual Perception:*** It is generally seen that vision is possibly the most dominant sense and our knowledge of things revolves around this. Even in the realm of VR, due significance is given to regeneration of this perception. A sense of depth is formed because of stereoscopic vision that permits the human eye to differentiate near and distant objects.

Along with this, several optical illusions such as parallax are also taken into consideration while designing a VR system. Complicated details of shadows and the extent of light that requires to be projected all fall under the same realm of visual perception. Critical Fusion Frequency refers to the rate and the number of static images that requires to be exposed so as to produce the illusion of a moving object. It is thus extremely significant to take into consideration all such factors while forming a visual perception for the user in a VR environment.

2. ***Sound Perception:*** In a real-life scenario, sound sources can be traced by our ears in any point in space. It becomes progressively challenging to execute the same in a VE due to the constraint of sound sources that can exist in the system. This can be addressed to a degree by using sounds of diverse frequencies and intensities to generate an illusion of a diverse sound source.
3. ***Touch Perception:*** It is imperative to create the perception of touch when the user comes in contact with an object in a VR System. It is also important to note that there are several parameters embraced in building a sense of touch; the force that should be feel by a user when he come in contact with the object and even the temperature of the object that a user feel while he or she is in contact. These can be attained through haptic feedback, where the intensity and frequency of the vibrations produce the illusion of contact for the user.
4. ***Olfactory Perception:*** The sense of how a scene would smell is representative and linked with our perception of the scene. We subconsciously connect our sense of smell with the sensitivity of a place. Thus, VR systems replicate the identical smells offering another dimension to the illusion of reality.

8.6. FUTURE CHALLENGES IN VIRTUAL REALITY (VR)

8.6.1. Hardware Challenges

One of the major challenges that come in the way of VR is to progress upon the tracking systems. It is well known that VR relied upon simulating several environments. Thus, it becomes essential to precisely track the several physical dynamics of the environments that are being re-created.

Another challenge is to look for the approach to naturalize the interaction between the VE and their users. There are several mass consumer technologies such as Microsoft Kinect that have engaged several motion-sensing technologies to keep an eye on the motion of a user with respect to the software.

One of the key obstructions is to precisely recreate Haptic perception which is the feeling felt by the skin. The visual presentations design in VEs is also intricate because the human visual system is extremely vulnerable to perceived anomalies, especially in motion scenes.

The majority of the technologies that are now using in the VR systems are originally designed for other disciplines. Therefore, there is slight movement in the world of technology to initiate a thrust towards concentrating on hardware primarily intended towards developing VR systems. Ergonomics is also a major concern when it comes to VR.

Hardware that are used in VR is generally clumsy and confine the physical freedom of the user. This would result in causing troublesome to the user's sense of inertia or balance. Bad ergonomics may also lead to a sense of cybersickness. It is often seen that VR technologies users often experience several levels of sickness ranging from nausea to headaches.

8.6.2. Ethical and Psychological Challenges

In addition to hardware challenges, there are psychological and ethical challenge facing VR too. Complete immersion in a VE could impact certain users psychologically. Generally, in VEs, users will be exposed to unethical and criminal acts.

Therefore, since the immersion level of VR is extremely high, there are high likely that the user may become de-sensitized to such actions. Another problem is the physical and emotional impact that might be the case in virtual stimuli.

The feelings of anxiety, fright, and nervousness will be intensified within an immersive VE in comparison to other forms of computing. Thus, there endures the risk of actual emotional trauma conducive to virtual incident.

8.7. VIRTUAL REALITY (VR) IN DIFFERENT INDUSTRIES

It has been seen that within the several years, VR has undergone marvelous improvement driven by the fact that it has a diversity of applications in several industries. Some applications for VR are discussed as follows:

- ***Scientific Visualizations:*** It is important to note that VR has incredible prospects in the area of scientific study where VR systems can be taken into usage to study complex structures and are envisioned and re-formed for further study. The Large Haudron Collider and CERN have established the system for scientific study.
- ***Medical Industry:*** The learning of anatomy, which had until now been limited to 3 D models, are being swapped by VR models that can be interesting and are comprehensive in nature. With the representing of the whole human anatomy, simulations, and the results of several medical experiments can be witnessed with exceptional detail. VR is definitely the tool for medical research of this century.
- ***Architectural Visualizations:*** These VR systems can map multifaceted architectural structures with comparative ease. They are interchanging conventional CAD (computer-aided design) models like the 3 D visualization can demonstrate much more detail. It also assists in architects that have been a follower of using 2D tools to visualize and create objects that exist in 3D.
- ***Education and Training:*** Severe changes in methods of education and training are spotted with the introduction of VR. High-quality content offered through stimulation of many more senses makes the information and data easier to comprehend and grasp. With respect to the training area, VR is widely been known to curtail down the costs by simulating real life scenarios and recreating VEs.
- ***Entertainment:*** Believed to be the next big revolution in this area, content delivery via VR guarantees a top class, immersive

experience. Sony has recently developed a VR headset for the next generation of gaming consoles that provide a far better experience in comparison to other gaming consoles already present in the market.

8.8. VR IN SOCIALIZING

It is worth noticing that there are already a number of VR-based social platforms that provide the opportunity to make new friends or allow strangers to meet up and play or chat in VEs, such as Altspace VR, VR Chat, and Rec Room (Figure 8.5).

Figure 8.5: Socializing in virtual reality.

Source: Image by Srushti IMX.

As with VR in other fields, the rising level of immersion that is conceivable thanks to new technological developments that will make them more attractive and more useful to mainstream audiences throughout the next few decades.

This year Facebook, which earlier had a large proportion of stake in VR because of its acquisition of headset manufacturer Oculus, revealed its Horizon platform. Presently, in beta, it provides the opportunity to people to erect and share collaborative online worlds where they can play games, hang out, or work together on cooperative projects.

While it is well known that people always find time to meet up with friends and loved ones in the real world, as our school and working lives become gradually remote, it is highly possible that the majority of our social interaction will move into the online realm, too.

Just as we are no longer barred from educational or career opportunities because of a progressively virtualized world, there would be more meaningful ways to unite with other humans as technology advances in this arena.

8.9. THE USE OF VR TO CHANGE THE FUTURE OF MUSIC

It is well known that music is an experience. Consumers and makers of music now stress more and more immersive as well as interactive experiences that necessitate the need for innovative designs, technologies, and models (Figure 8.6).

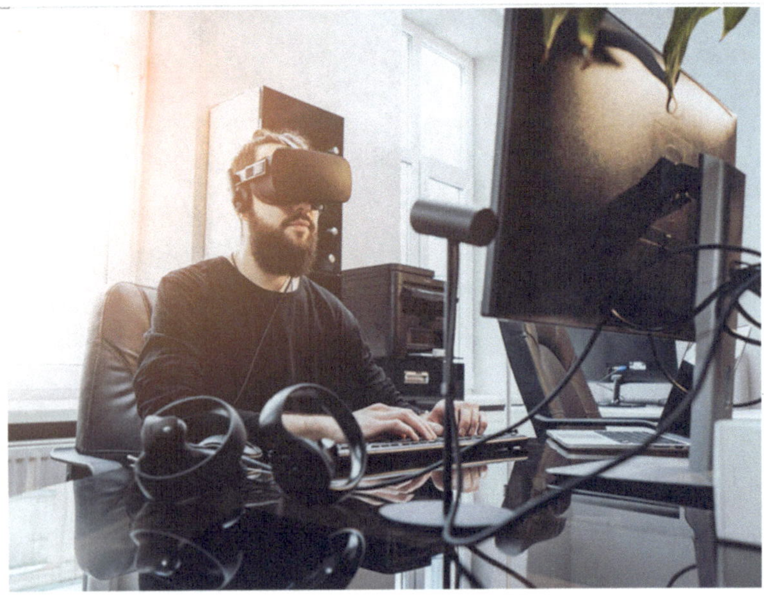

Figure 8.6: Virtual reality in music editing.

Source: Image by MusicTech.

These revolutions can take on many forms, embracing offline services, online programs, and physical products, but in this section, the main area of focus will be only on those of the mobile digital features that prevailing or future mobile applications could ponder implementing based on technology such as VR and AR.

Some of the trends that is currently in existence to music streaming and related technologies are growing because of the demands of mediated-and-live music. While live music is borne to its situational real-time presentation context and its engaging elements, mediated music denotes more usually to the channels and means of its distribution, through which sounds are conveyed over various media, normally not in real-time, and usually with substantial production processing.

The newer genre of mediated-and-live music seeks to assimilate both benefits concurrently, reaching a large number of younger audiences, while at the same time ensuring an immersive and interactive experience.

This trend advances itself better to newer technological integration, such as VR and AR, which can improve the overall experience of music by aiming music consumers who crave the "best of both worlds" (*live and mediated*). It is often seen that nowadays social media is used for music streaming.

Mediated streaming of festival music surges during the festival time; live festivals push the flow of audiences to use streaming services and social media, and post-hoc evaluation produces ongoing revenue streams to taped music (Danielsen and Kjus, 2019).

8.9.1. Virtual Reality (VR) Designs for Music

It is important to note that the potential of VR to entirely alter the way music is both created and consumed is huge. By simulating intense concert environments, or putting users in a recording studio, VR can offer somebody with a fully immersive music experience from their living room.

The first project planned is a complimentary concert service. If, according to Predictive Coding, human experience is an internal simulation of the body as well as near environment, a properly executed VR experience of a concert might offer somebody with the similar feeling like the one who is in the front row at their much-loved artist's concert from anywhere in the world.

It was demonstrated that 360° cameras and sensors could be placed in the front row of live concerts, providing the opportunity for consumers to live-stream the concert with a VR headset and experience the feeling of the same on a real time basis. This, of course, would be the case only after getting permission from the team of artists whose concert is being streamed.

Why would any artist would like to lessen the number of people from their crowd and back to their living rooms? This experience would definitely not be free, but only accessible to ticket holders, comparable to the live concert itself.

Take Billy Joel, whose concert at the Casino and Hard Rock Hotel and in Hollywood, FL on January 2020, sold front row tickets for $811. Apart from just selling a meager number of these front row tickets, additional front row VR tickets could be offered at a comparative much cheaper price, say for as low as $75.

This would generate extra revenue for the Billy Joel as well as Hard Rock Hotel while providing the opportunity to fans who may not have the money or the time to attend the concert on a real time basis to still get an authentic experience.

To make the experience more social and authentic, everyone who has the authorization to VE of live concert will also give access to do live chats with other participants in a live virtual chat room.

With the recent coronavirus pandemic, it is high likely that people have to become used to this trend and enjoy them even in a similar manner, one could also imagine that this could be the future of immersive music experience for even more reasons.

The notions above could be combined with AR technology for a higher immersive experience. Apart from attending the concert in VR, AR potentials could offer animated effects such as fluctuations in weather conditions, fireworks going off on stage, or even alterations in the artist on stage themselves. Let's taken into consideration Travis Scott concerts as an example.

Travis Scott is widely known for constructing amusement park rides on stage, such a roller coaster that stretches over the crowd. The VR ticket holders at the Travis Scott concert might not be able to only watch roller coaster, but at the same time could also experience it, as well as a wholly AR-generated amusement park in full swing neighboring the whole concert.

Finally, VR technology could be exploited for music creation. Instead of just acquiring high-cost expensive instruments, speakers, as well as other equipment needed to record music, a VR program could pretend this experience through a virtual recording studio.

Comparable to the VR games stated above, such as Rock Band, where players offered with toy instruments, a VR recording studio could also provide the opportunities to someone aiming to create its own music. This type of technology could entirely alter the way music is made. Thus, it is helpful for both the creator as well as listeners of music. In addition, it also able to attract those audiences that do not have an interest in music.

8.10. THE FUTURE OF PAYMENTS: TRENDS AND OPPORTUNITIES OF AR AND VR

How the adoption of immersive technologies could transform the way the payment industry performs?

The applications of AR and VR are no secret for businesses across diverse industry verticals. From virtual gaming to transforming future workplaces, these immersive technologies have a considerable impact on day-to-day business activities. The financial services industry is not an exemption as these technologies assures more prospects to financial institutions and payment companies to involve their customer in a virtual world.

VR and AR have already taken FinTech by storm, facilitating companies and startups to examine and take apt actions for quickly altering customer behaviors. According to the Goldman Sachs,' one of the biggest global investment banks, it was estimated that virtual and augmented realities are likely to become an 80-billion-dollar market by the year 2025.

Financial institutions are testing with these technologies by emphasizing on the increased value propositions that can be originated by making the experience more accessible and personalized core banking systems.

Both AR and VR are normally redefining the way financial and banks services providers bring more personal banking experience to their clients and customers. For instance, Fidelity Investments brought a VR agent, named Cora, which has the potential to interact with customers by endorsing stocks, evaluating the performance of a company, pulling up financial charts, and decide whether it will be vice to put it into the investors' portfolio.

Some financial institutions are leveraging virtual and AR to make trading a virtual experience by creating more and more virtual workstations.

For instance, financial services firm Citi uses Microsoft HoloLens to offer traders Holographic Workstations. This provides 2D and 3D elements which combines with the existing process of the bank. On the other hand,

Comarch leverages VR in its wealth management software to offer users better access to trading tools and algorithms.

8.10.1. Heightened Payment Experience with Virtual Retail Banking

These days, the financial industry has been growing at a rapid rate towards the digitization, and virtual retail banking concept is begun rolling out. For instance, The Hong Kong Monetary Authority (HKMA), recently revealed 8 new virtual banking licenses to operators that will introduce new digital-only banks.

With the help of virtual assistants and chatbots, this banking experience is becoming more widespread, delivering immersive and personalized customer experience.

It is generally seen that with the help of virtual banking, one can simply provide a face-to-face interaction from any time to person living at anywhere and gives key-value addition that most financial services firms are seeking with AR, simplifying better interaction and communication, in addition to gaining trust among customers.

Furthermore, with the help of AR application, one can provide customers with information with respect to payments and ensures how their funds are conveyed to another account. VR and AR in banking can also be taken into usage to create a more secure customer experience.

For instance, biometric security could be introduced in an AR system, consequently, associate with a VR world. These can be fully utilized to access several banking processes such as ATM transactions, VR bank services, and payments.

According to several market reports and research surveys, the VR market is projected to grow at a CAGR of 33.5% from US$7.9 billion in 2018 to US$44.7 billion by the year 2024.

8.11. FOUR WAYS VR WILL IMPACT THE FUTURE OF TECH CAREERS

Forget checking online shopping or social media, because it's thinkable (and probable), that the future of lunch breaks will comprises of strapping on a headset to ride an underwater rollercoaster. (And you thought you couldn't spice up your workday!) (Figure 8.7).

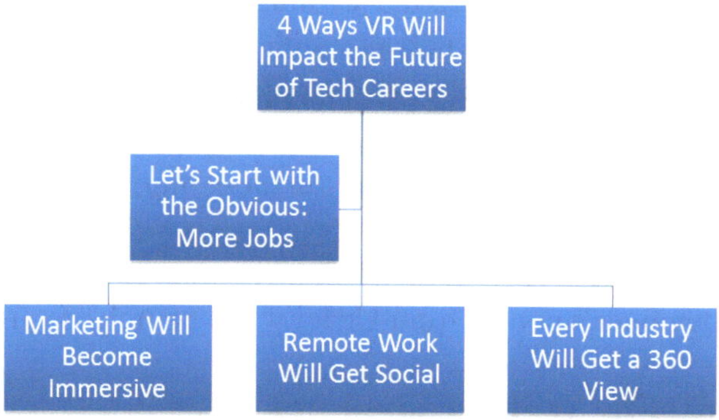

Figure 8.7: Ways in which VR can impact the future of tech careers.

VR is not just for play, though, and one cannot doubt that this technology will completely transform the way how one work in their day-to-day life. This is particular true for various developers as well as other technical folks, but the VR evolution will likely to have an impact on the entire industry.

A little context here—VR, while still considered to be in its infancy, is growing at a rapid rate. According to Deloitte Global, it was estimated that 2.5 million VR headsets were sold worldwide last year. And though the technology has not created its space in every household, it is highly likely that 2017 is the year that VR goes mainstream.

And this expanding industry is more likely to completely transform that way how people do jobs by making it more exciting. Here's how.

8.11.1. Let Us Start with the Obvious: More Jobs

It is worth noticing that there is no need to be a hardcore gamer in order to understand that VR's influence on the video game industry is huge. And with the gradual growth in industry, development jobs in the field of VR will also grow in the same proportion.

The best part? There is no need to be a game developer to explore opportunities, as the industry will require support from those having experience in UX/UI, marketing, legal, business development, and more.

Sound interesting? If one think that they might need to get into VR gaming, it's the time to start training now. There are meetups and conferences happening all over the country, such as the Experiential Technology Conference, the Game Developers Conference, the IEEE VR Conference in Los Angeles. + Expo in San Francisco.

It is always advisable to keep the learning on, and looking for the people that are already in the space to network with along the way.

There is also a need of active participation from your side to get closely familiar with VR technology. Look for a course that lets you to hone skills in technologies that VR companies care about modeling, 3D animation, object-oriented programming, and software development kits (SDKs). There are a large number of pertinent programming courses out there- and some of them, like a beginner's course from Udacity, are completely free.

8.11.2. Marketing Will Become Immersive

Marketing agencies are basically the first one that try their hands on the new digital technology, and VR is no exception. Gareth Price, who is the Technical Director at New York-based digital agency Ready Set Rocket, said that "We are definitely watching VR trends with a keen eye."

Agencies similar to this one is building relationships with specialists who can produce VR content for digital marketing campaigns. Price believes such sorts of tech jobs "will blend the skills needed of developers, designers, 3D artists and filmmakers, in new and more cooperative ways."

His advice to employees of digital agency is too aware or get a full understanding of this technology as early as possible. Ready Set Rocket has Oculus and Samsung rigs in-house for analysis purposes, and Price inspires his staff to take them home for a spin. "Start testing on your own time right now as an investment into your career," he says. "Today's side project will turn into future high-flying job."

8.11.3. Remote Work Will Get Social

Since Oculus was acquired by Facebook in the year 2014, it has been exploring various ways to integrate VR into the social networking experience. When CEO Mark Zuckerberg staged at the Oculus Connect conference, he demonstrated that he was able to cooperate with participating co-workers in an immersive virtual world. He also states that just how revolutionize the technology might be for the way you work in the future.

So, if a person is frustrated of unproductive work calls, take heart: one might be soon able to initiate its interviews, brainstorming sessions, and status updates from a beach in Mexico.

Having a virtual office, one might be able to connect with colleagues stationed across the world for more effective and personal meetings. Now, there is a time to brush up on soft skills like being a good collaborator!

8.11.4. Every Industry Will Get a 360 View

If a person is not in tech, then it does not mean VR will not impact its career. Research from design, product strategy, and development company Yeti shows that close to 54% of U.S. product developers are already actively participating in VR projects, with applications varying from entertainment to education, travel, and healthcare.

According to Gareth Price, one of the major challenges that companies are facing with VR right now is figuring out "how to tell stories well [using VR]," so brush up on your writing, plotting, and storyboarding skills, or having knowledge about brands that are already doing it well.

Once the company will find what they want to demonstrate to their customers—say, the importance of your hospitality brand's swim-up bars—one will be likely to watch how VR could improve it. All that's left is to find a development partner to bring your brand story to life.

With a plethora of companies testing with this technology, it won't take much time before VR is a reality in jobs of all kinds.

8.12. AR AND VR: SOME KEY PLAYERS

8.12.1. Facebook

Some of the years back, Mark Zuckerberg has told that he has an aim to make 1 billion people to try VR. It's still has a long way to go to make this actually happen. Facebook acquired Oculus company, which it bought for around $2 billion, offers three headsets: Oculus Quest for VR gaming, the all-in-one Oculus Go for VR viewing, and the Oculus Rift S, which requires to be associated to a PC, for high performance gaming.

Facebook also acquire CTRL-Labs that would also help do away with the requirement for controllers in VR. "I don't believe that it's a thing of 2021. But optimistically, it's not a 2030 thing," said Zuckerberg when asked when VR will hit the mainstream.

8.12.2. Apple

There are a variety of AR apps that are currently in existence for the iPhone, using its ARKit developer tools. These apps provide the opportunity to users to see virtual elements added to the view they can see through the smartphone screen.

Ikea's Place app permits users to plunge virtual furniture and other home furnishings into a room to watch how they would appear before buying them.

Beyond this, it also came to notice that Apple has secretly working on some form of AR headset. "I'm extremely thrilled by AR because I can see uses for it everywhere," Apple CEO Tim Cook said back in 2017.

8.12.3. Microsoft

Microsoft's HoloLens is a self-contained 'mixed reality (MR)' headset enabled with the facility of wi-fi, and is intended for the use in companies and businesses. Some of the early users of this are Volvo and Lift company ThyssenKrupp. Microsoft is aiming at to launch Hololens 2 whose costs would be around $3,500.

8.12.4. Google

It is worth noticing that Google has for so long been fervent about AR and endures to pursue it in various directions. Google Glass is till now considered as the best-known attempt at building a headset: while the consumer version met with ridicule, the Glass Enterprise Edition has been used by GE, DHL, and Sutter Health to make staff more creative.

Within Android, Google has ARCore, a platform that is used for building AR apps for its own Pixel smartphones, as well as handsets from Huawei, Samsung, and others.

It was also found that Google is also adding more AR features to its own apps. It just added Live View to Google Maps so that when you can hold up your phone and see direction and arrows overlaid on the view of the streets in front of you.

"Looking ahead, I believe that we are entering a new phase of computing: an era of the camera, if you will," said Aparna Chennapragada VP, Google Lens and AR last year.

8.12.5. Magic Leap

Magic Leap defines its headset as a 'spatial computer' that permits or provides the opportunity to interact with digital content. "So, when you make digital penguins walk off the edge of the coffee table, they fall off the edge of the coffee table, just like real penguins would," it says. The headset that are necessitate the need to carry a 'lightpack' that carries out the processing, sells is sold for the cost of $2,295.

8.12.6. HTC

HTC has just added to its portfolio Vive range of headsets with the HTC Vive Cosmos. It basically makes use of a group of six cameras to monitor the location of the headset and its user, instead of the laser-emitting base stations which were used earlier. Recently, it came to notice that HTC is working with the BBC on an immersive doctor who experiences.

8.13. CONCLUSION

In the end, it is concluded that VR has wider prospects than it seems to look like. In the future, one can expect VR to be applicable in various fields, even in that that is beyond imagination. In the present scenario, the use of VR has been continuously increasing, and in the future, if this rate continues, it would be adopted by the masses. With the advancement in technology and high speed of internet, the development in this field is also expected.

One can see that several big players such as Google, Facebook, Microsoft are entering into this market, it is possible only because they are seeing some future in this, they have employed dedicated team for this that is focusing on innovations and research and development in VR. So, one can expect some revolutionary change in the field of VR in the near future. In addition to it, several tech companies are also employing VR technology into their companies to provide training and development to their employees in order to extract the maximum benefits.

However, there are several areas of improvement that should be taken into consideration if one is really concern about future VR prospects. It is a continuous process that will endure till the life of VR.

REFERENCES

1. (2019). *What is the Future of Augmented and Virtual Reality Technology?* [eBook] Available at: https://www.jabil.com/blog/future-of-augmented-and-virtual-reality-technology.html (accessed on 29 April 2021).
2. BOSS Magazine, (2021). *BOSS Magazine | The Future of Virtual Reality*. [Online] Available at: https://thebossmagazine.com/future-virtual-reality/ (accessed on 29 April 2021).
3. Clementson, M., (2021). *The Future is Virtual Reality: Using VR Technology in Architecture*. [Online] IBI Group. Available at: https://www.ibigroup.com/ibi-insights/future-virtual-reality/ (accessed on 29 April 2021).
4. Dialani, P., (2021). *Future of Communication: Virtual Reality is Stepping Up*. [Online] Analytics Insight. Available at: https://www.analyticsinsight.net/future-of-communication-virtual-reality-is-stepping-up/ (accessed on 29 April 2021).
5. Gobbetti, E., (2021). *Virtual reality: Past, Present and Future*. [Online] www.researchgate.net. Available at: https://www.researchgate.net/publication/12947828_Virtual_reality_Past_present_and_future (accessed on 29 April 2021).
6. Harvard Business Review, (2021). *Is VR the Future of Corporate Training?* [Online] Available at: https://hbr.org/2020/09/is-vr-the-future-of-corporate-training (accessed on 29 April 2021).
7. Hniedkova, O., Osipova, N., Kravtsov, H., & Lishchuk, T., (2020). *Technologies of Virtual and Augmented Reality for High Education and Secondary School*. [eBook] Available at: http://ceur-ws.org/Vol-2393/paper_258.pdf (accessed on 29 April 2021).
8. Iberdrola, (2021). *Virtual Reality: Another World Within Sight*. [Online] Available at: https://www.iberdrola.com/innovation/virtual-reality (accessed on 29 April 2021).
9. Information Age, (2021). *The Future of Virtual Reality-Information Age*. [Online] Available at: https://www.information-age.com/future-virtual-reality-123461990/ (accessed on 29 April 2021).
10. Kumar, V., (2021). *The Future of Payments: Trends and Opportunities of AR and VR*. [Online] IndustryWired. Available at: https://industrywired.com/the-future-of-payments-trends-and-opportunities-of-ar-and-vr/ (accessed on 29 April 2021).

11. Lindbergh, B., (2021). *Waiting for the Future of Virtual Reality.* [Online] The Ringer. Available at: https://www.theringer.com/2021/1/12/22226387/virtual-reality-playstation-xbox-oculus (accessed on 29 April 2021).
12. Lynch, M., (2021). *What is the Future of Virtual Reality? - The Tech Edvocate.* [Online] The Tech Edvocate. Available at: https://www.thetechedvocate.org/what-is-the-future-of-virtual-reality/ (accessed on 29 April 2021).
13. Marr, B., (2021). *The Future of Virtual Reality (VR).* [Online] www.forbes.com. Available at: hhttps://www.forbes.com/sites/bernardmarr/2020/12/18/the-future-of-virtual-reality-vr/?sh=212df44527be (accessed on 29 April 2021).
14. McCloy, R., & Stone, R., (2001). Science, medicine, and the future: Virtual reality in surgery. *BMJ., 323*(7318), 912–915.
15. Medium, (2019). *Why Virtual Reality is the Future?* [Online] Available at: https://medium.com/predict/why-virtual-reality-is-the-future-bdc470ecee28 (accessed on 29 April 2021).
16. Mike Gingerich, (2021). *The Future of Virtual Reality and Business-Mike Gingerich.* [Online] Available at: https://www.mikegingerich.com/blog/future-virtual-reality-business/ (accessed on 29 April 2021).
17. Raghavan, R., & Rao, P., (2020). *Immersive Learning for the Future Workforce.* [eBook] Available at: https://www.accenture.com/_acnmedia/pdf-86/accenture-extended-reality-immersive-training.pdf (accessed on 29 April 2021).
18. Rahman, T., (2020). *The Future of Virtual Reality.* [Online] Medium. Available at: https://medium.com/predict/the-future-of-virtual-reality-1880dce95f59 (accessed on 29 April 2021).
19. Scholars, W., Decesare, M., & Wang, H., (2021). *The Use of AR and VR to Change the Future of Music-Wharton Neuroscience Initiative.* [Online] Wharton Neuroscience Initiative. Available at: https://neuro.wharton.upenn.edu/community/winss_scholar_article4/https://neuro.wharton.upenn.edu/community/winss_scholar_article4/ (accessed on 29 April 2021).
20. TechPats, (2020). *The Future of VR | How Viable is it?* [Online] Available at: https://www.techpats.com/future-worlds-virtual-reality/ (accessed on 29 April 2021).

21. Terrier, D., (2021). *Virtual Reality: The Future of Education?* [Online] Hospitalityinsights.ehl.edu. Available at: https://hospitalityinsights.ehl.edu/virtual-reality-future-of-education (accessed on 29 April 2021).
22. Toptal Insights Blog, (2017). *Virtual Reality Applications Drive the Future of Work*. [Online] Available at: https://www.toptal.com/insights/future-of-work/virtual-reality-applications-future-work (accessed on 29 April 2021).
23. Toshniwal, R., & Ghosh, D. K., (2014). Virtual reality: The future interface of technology. [eBook] *International Journal of Computer Science and Information Technologies*, Available at: http://ijcsit.com/docs/Volume%205/vol5issue06/ijcsit2014050630.pdf (accessed on 29 April 2021).
24. Wegert, T., (2021). *Four Ways VR Will Impact the Future of Tech Careers*. [Online] www.themuse.com. Available at: https://www.themuse.com/advice/4-ways-vr-will-impact-the-future-of-tech-careers (accessed on 29 April 2021).
25. Y-sbm.com. (2021). *The Future of VR Technology: Six VR Applications you Need to Know*. [Online] Available at: https://y-sbm.com/blog/future-of-virtual-reality (accessed on 29 April 2021).

Index